T0140841

Predictive Control under Uncertainty

From Conceptual Aspects to Computational Approaches

Von der Fakultät Konstruktions-, Produktions- und Fahrzeugtechnik
und dem Stuttgarter Zentrum für Simulationswissenschaft
der Universität Stuttgart zur Erlangung der Würde eines
Doktor-Ingenieurs (Dr.-Ing.) genehmigte Abhandlung

Vorgelegt von

Matthias Lorenzen

aus Herrenberg

Hauptberichter: Prof. Dr.-Ing. Frank Allgöwer
Mitberichter: Prof. Maria Prandini, Ph.D.
Prof. Dr. sc. Melanie Zeilinger

Tag der mündlichen Prüfung: 28. November 2018

Institut für Systemtheorie und Regelungstechnik
der Universität Stuttgart

2019

Bibliografische Information der Deutschen Nationalbibliothek

Die Deutsche Nationalbibliothek verzeichnet diese Publikation in der Deutschen Nationalbibliografie; detaillierte bibliografische Daten sind im Internet über http://dnb.d-nb.de abrufbar.

D93

ISBN 978-3-8325-4980-0

Logos Verlag Berlin GmbH
Comeniushof, Gubener Str. 47,
10243 Berlin
Tel.: +49 (0)30 42 85 10 90
Fax: +49 (0)30 42 85 10 92
INTERNET: https://www.logos-verlag.de

Acknowledgments

With the completion of this thesis, a time that was professionally as well as personally instructive has come to an end. A time in which I could experience the ups and downs of scientific work, which taught me to stay focused after hours of trying to grasp the essence of an article, but also to take a break when being stuck in a potential dead end.

I gratefully acknowledge all those who made this possible. First and foremost, I want to express my gratitude to Prof. Frank Allgöwer for his guidance and support, as well as for granting extraordinary scientific freedom. At the Institute for Systems Theory and Automatic Control (IST), he successfully created a stimulating and cosmopolitan environment, which motivated me to read and learn a lot and which provided numerous possibilities to discuss and collaborate with inspiring people.

I am indebted to the late Prof. Roberto Tempo and Dr. Fabrizio Dabbene, who became invaluable scientific mentors. Both introduced me to the topic of finite sample approximations in stochastic programming and regularly hosted me in Turin. I greatly benefited from the close collaboration, their advice, and the feedback they provided. Furthermore, I want to express my gratitude to Prof. Mark Cannon, who was a great host during the three months I have spent at the University of Oxford and who has been a role model for rigor in the approach to Stochastic MPC. I enjoyed the inspiring atmosphere and learned a lot from the scientific discussions we had and which led to fruitful research results.

Special thanks go to the members of my doctoral examination committee, Prof. Maria Prandini, Prof. Melanie Zeilinger, and Prof. Alexander Verl, for taking the time, their interest in my work, and their encouraging comments.

Aside from the faculty, I acknowledge the support of all my colleagues at the IST. In particular, I want to thank Florian Bayer and Matthias Müller for the close collaboration and Meriem Gharbi, Julian Berberich, Johannes Köhler, Philipp Köhler, Matthias Müller, and Raffaele Soloperto for proofreading this thesis and their helpful comments. I am indebted to Martina Mammarella for applying the control algorithms in an experimental setup.

Finally, I am grateful to all my friends who helped me clear my mind while climbing, cycling, or listening to some live music over a beer. In particular, I want to thank Steffen Raach, for our inspiring and joyful weekly meetings and the faithful friendship. Last but not least, I want to express my eternal gratitude to my family and my partner Anja for their unconditional support, patience, and positiveness.

Table of Contents

Abstract

This thesis addresses constrained control of systems under stochastic disturbances and model uncertainty. In particular, we study stochastic and adaptive model predictive control (MPC) algorithms to solve such problems. Given a stochastic model, the focus is put on chance constraints, tractable approximations, and their implication on feasibility of the online optimization in a receding horizon framework. Model uncertainty is addressed by online parameter identification to reduce conservatism and improve closed-loop performance.

More specifically, building upon the analysis of conceptual aspects, we develop computational approaches for linear systems with additive and multiplicative disturbances. The analysis of a non-conservative, computationally tractable relaxation of chance constraints leads to an important separation of sufficient conditions for feasibility and stability. The latter is of particular interest for rigorously applying finite sample approximations to solve the online stochastic optimal control problem. We discuss the differences between online and offline sampling and provide explicit bounds on the sample complexity to guarantee satisfaction of chance constraints with a user chosen confidence. The proposed algorithms provide rigorous guarantees for relevant properties like feasibility of the online optimization, constraint satisfaction, and convergence of the closed-loop system.

For linear systems with model uncertainty, we propose a computationally tractable framework for MPC with online parameter identification. Recursive feasibility and constraint satisfaction is guaranteed through the combination of set-membership system identification with a prediction tube. Additionally, to prove finite gain stability for the closed loop, a suitably chosen point estimate of the uncertain parameters is employed in a certainty equivalence approach.

Finally, we address stochastic MPC without terminal constraints and derive sufficient conditions to establish closed-loop stability. Unlike the results for MPC with terminal constraints and cost, the stability proof does not directly extend to the discussed computational approaches, but necessitates stronger assumptions. These are analyzed for general nonlinear systems along with examples and simplifications for linear systems.

Deutsche Kurzfassung

Diese Arbeit befasst sich mit der Regelung dynamischer Systeme unter Berücksichtigung von Beschränkungen sowie stochastisch modellierten Störungen und Modellunsicherheiten. Insbesondere wird modellbasierte prädiktive Regelung zur Lösung solcher Problemstellungen untersucht. Bei der stochastischen Modellierung liegt der Fokus auf Beschränkungen, welche mit einer vorgegebenen Wahrscheinlichkeit eingehalten werden sollen sowie praktisch umsetzbaren Annäherungen und die daraus resultierende Auswirkung auf die Lösbarkeit des Optimalsteuerungsproblems. Zur Reduzierung von Modellunsicherheiten wird eine zur Laufzeit des Algorithmus durchgeführte Parameteridentifikation betrachtet.

Auf einer Analyse theoretischer Aspekte aufbauend, werden praktisch umsetzbare Algorithmen für lineare Systeme mit additiven und multiplikativen Störungen entworfen. Die Untersuchung einer Relaxation der probabilistischen Beschränkungen führt zu einer neuen, separaten Betrachtung hinreichender Bedingungen für rekursive Lösbarkeit und Stabilität. Diese ist insbesondere bei der näherungsweisen Lösung des stochastischen Optimalsteuerungsproblems mittels einer endlichen Anzahl an Realisierung der Störung von Interesse. Dabei wird auf den Unterschied zwischen einer solchen Approximation zur Laufzeit im Gegensatz zum einmaligen Entwurf vorab eingegangen. Explizite Schranken an die notwendige Anzahl an Realisierungen, um die Einhaltung der Beschränkungen mit einer gewünschten Wahrscheinlichkeit zu gewährleisten, werden angegeben. Für die vorgestellten Algorithmen werden relevante Eigenschaften wie beispielsweise Lösbarkeit des Optimalsteuerungsproblems, Einhaltung gegebener Beschränkungen sowie Stabilität des geschlossenen Kreises bewiesen.

Für lineare Systeme mit Modellunsicherheiten wird eine mengenbasierte Parameterschätzung und Zustandsprädiktion vorgeschlagen, welche rekursive Lösbarkeit und die Einhaltung von Beschränkungen garantiert. Zusätzlich wird ein Punktschätzer der unbekannten Parameter entworfen, womit eine endlichen Verstärkung durch den geschlossenen Kreis bewiesen werden kann.

In Kapitel 5 wird die Annahme einer wohl gewählten Endbeschränkung und eines Endkostenterms fallen gelassen und alternative hinreichende Bedingungen für Stabilität untersucht. Da sich die Beweise eines konzeptionellen Algorithmus hier nicht direkt auf praktisch umsetzbare Approximationen erweitern lassen, sind stärkere Annahmen notwendig. Diese werden für allgemeine, nichtlinear Systeme analysiert, in Beispielen illustriert und für lineare Systeme vereinfacht.

Chapter 1

Introduction

Finding not only any, but finding the optimal solution for a given problem and objective is one of the most basic questions in the various fields of the exact sciences. Early, well known examples include Dido's problem of maximizing the area enclosed by a curve of fixed length, the brachistochrone problem posed by Bernoulli in 1696, or Hamilton's principle of least action which describes the equations of motion of a mechanical system.

All these classical examples have in common that they can be formulated and solved in the mathematical framework of the calculus of variations, formally developed by Lagrange and Euler between 1754 and 1756. In engineering, more specifically in systems theory and control, calculus of variations has led to the highly successful theory of optimal control, which studies the same problems, but from a viewpoint of dynamical systems with inputs, states, and outputs. Compared to the conceptual aspects of the variational methods, the maximum principle, developed in the seminal work of Pontryagin (1959), had a huge impact not least due to providing a practically applicable approach of computing optimal inputs for systems with arbitrary dynamics and constraints. At the same time, Bellman (1957) introduced the conceptual framework of dynamic programming for solving optimal feedback control and decision problems. Although the Hamilton-Jacobi-Bellman partial differential equation provides a necessary and sufficient condition for optimality and has led to computational approaches for solving linear optimal control problems, its benefits are difficult to achieve in full generality, in particular for the important cases of nonlinear or constrained systems.

Among the different computational approaches to approximate the dynamic programming solution, a highly successful and widely accepted method is model predictive control (MPC), often also referred to as receding horizon control. The basic idea, depicted in Figure 1.1, is to derive a local approximation of the optimal value function by continuously measuring the current process state and subsequently computing an optimal future control sequence. Feedback is obtained by applying this optimal input trajectory only during a short time interval, after which the process is repeated based on a new measurement.

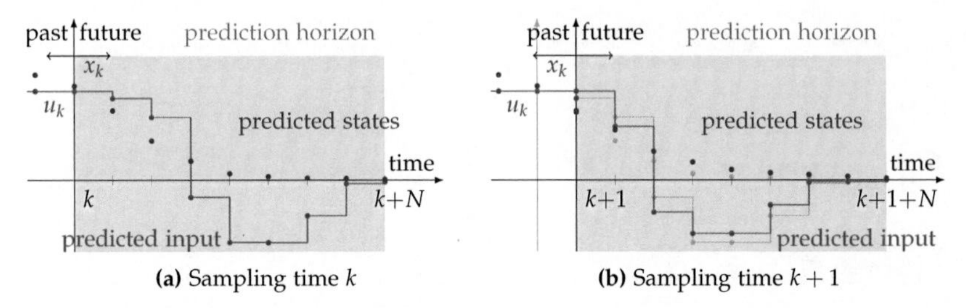

(a) Sampling time k **(b)** Sampling time $k+1$

Figure 1.1. Basic idea of model predictive control: Finite horizon optimal state and input trajectories given the measured state at time k and with shifted horizon given the measured state at time $k+1$.

Since the appearance of the conceptual idea in academia (Propoi, 1963; Dreyfus, 1965) and in particular the pioneering computational approaches and industrial application (Richalet et al., 1978; Cutler and Ramaker, 1979), it received significant and continuing attention from researchers with both, a theoretic as well as an application-oriented point of view. One of the main advantages of MPC, compared to most classical control strategies, is that it provides a unified and constructive approach to controller design for multivariable systems while explicitly incorporating state and input constraints, as well as a user-defined performance criterion. A particularly suitable and relevant example bringing together all aspects is that of fully automated flight. The task can be formulated as feedback control of a nonlinear multivariable dynamical system, where the performance criterion includes passenger comfort or fuel consumption and the constraints, given by the flight envelope and actuator limits, are safety critical and need to be adhered to under all circumstances.

Although disturbances and uncertainties, the mismatch between the real world and a mathematical model, can be considered one if not the *raison d'être* of feedback control, they are often treated only indirectly in MPC, namely through the repeated state measurement and solution of the optimal control problem. Dreyfus (1965) already noted that in presence of uncertainty, MPC "results in a control scheme that is inferior to true optimal feedback control". To cope with this disadvantage, in recent years the inherent robustness properties, as well as constructive methods to explicitly incorporate disturbance and uncertainty models into the MPC design received a great deal of attention. While unknown but bounded disturbance models allow robust approaches, often some statistical regularity can be observed and disturbances can be modeled by random variables. Having always been an integral part of the dynamic programming literature, it is only in recent years that stochastic models received major attention in the MPC literature.

The advantage of incorporating statistical information in the MPC design is that it allows to optimize the average performance or appropriate risk measures. Furthermore, it introduces the possibility to constructively include soft constraints, so called chance constraints, which can be violated up to a user-defined threshold. Similarly, if the uncertainties are known to be constant over time, performance can be increased by updating the model based on closed-loop measurements. Revisiting the example above, uncertainties like the total load or changes due to tear and wear can be estimated online in order to update the internal model. For disturbances like wind gusts, there exist accepted stochastic descriptions, which allow to incorporate chance constraints, for instance on the maximal acceleration to increase passenger comfort or decrease fatigue damage, which cannot be described by hard constraints.

The goal of this thesis is to develop a deepened system theoretic understanding of as well as computational approaches for predictive control of dynamical systems with uncertainties or disturbances modeled by a random process.

In the following two sections, we proceed with a review of relevant, related literature and conclude by summarizing the contributions of this thesis.

1.1 Research topic overview

Model predictive control

Much research effort has been devoted to understand the conceptual aspects of stabilizing model predictive control and different sufficient conditions for closed-loop stability have been derived, which by now achieved textbook maturity. Detailed results for discrete-time systems are summarized in (Rawlings et al., 2017) with a focus on MPC algorithms with terminal constraint and terminal cost. Nonlinear discrete-time systems, implementation details, and in particular sufficient conditions for stability without terminal constraint are rigorously presented in (Grüne and Pannek, 2017) with a similar stability analysis for continuous time in (Reble, 2013). In (Camacho and Alba, 2007) emphasis is put on application and most recent results on robust and stochastic MPC, although mainly for linear systems, are published in (Kouvaritakis and Cannon, 2016). Despite different analysis methods, most approaches have in common that to prove stability, the optimal value function is carefully designed in order to be used as Lyapunov function.

In the following, we focus on the most relevant literature directly related to the topic of this thesis. For a more comprehensive review of the different MPC schemes for various system classes and control tasks, as well as application examples the interested reader is referred to the survey papers (Mayne et al., 2000; Mayne, 2014; Qin and Badgwell, 2003; Darby and Nikolaou, 2012).

Stochastic MPC Since robust approaches to MPC for linear models, e.g., (Kothare et al., 1996; Langson et al., 2004; Chisci et al., 2001; Mayne et al., 2005) as well as for non-linear models, e.g., (Magni et al., 2001; Limon et al., 2009) are often either too conservative or remain mainly on a conceptual level, over the last two decades, stochastic MPC, where disturbances and uncertainties are modeled by random variables, received significant attention from both practical and theoretically-oriented researchers.

As with the general development of MPC, first publications mainly addressed *computational approaches*, in particular methods to propagate the uncertainty for efficiently evaluating the cost function and the chance constraints. In the pioneering work of Schwarm and Nikolaou (1999), motivated by the lack of robust constraint satisfaction in nominal MPC, the system is described by an FIR model with exogenous disturbance and coefficients described by random variables. The authors introduce the concept of chance constraints into MPC and provide a computational tractable reformulation as second order cone constraints. In (Zhang et al., 2002) optimal steady state calculation in MPC for constrained systems with stochastic uncertainty in the plant model is developed and its economical advantage is highlighted in a small case study. Additionally, in (van Hessem et al., 2001; van Hessem, 2004) feedback gains are optimized to shape the state distribution and it is shown how the online optimization can be convexified through Youla parametrization or innovation feedback. Based thereon, an extension to nonlinear systems through sequential conic programming has been presented in (van Hessem and Bosgra, 2006). A notable exception to most publications is (Batina, 2004), where a dynamic programming like backward recursion to determine an optimal feedback solution and sampling to approximate the expected value is proposed.

By now, different methods to evaluate exactly, approximate, or bound the desired quantities have been proposed in the stochastic MPC literature. An exact evaluation is in general only possible in a linear setup with Gaussian noise through a conic programming (van Hessem, 2004) or with finitely supported disturbances as in (de la Peña et al., 2005; Bernardini and Bemporad, 2012) through applying multi-stage stochastic optimization based on exhaustive enumeration of scenarios. Approximations include the so called "particle approach" (Blackmore et al., 2010) or "scenario tree" (Lucia et al., 2013), where the stochastic program is approximated by using possibly weighted samples of the random variables, but no guarantees or rigorous guidelines on choosing the number of samples are provided. Mesbah et al. (2014) and Paulson et al. (2014) approximate the stochastic program with polynomial chaos expansion, which is mainly restricted to systems with a small number of random variables, but can be applied to non-linear models. Conservativeness in the approximations has been addressed in (Korda et al., 2014) through an online adaptation based on observed disturbances. Deterministic bounding methods have been derived mainly from general tail inequalities, like the Cantelli inequality used in (Farina and Scattolini, 2016). In recent years, randomized algorithms (Tempo et al., 2013) for controller design received significant attention. In the context of MPC, scenario theory provides confidence bounds on the satisfaction of chance constraints and has been employed to determine an optimal feedback gain (Kanev and Verhaegen, 2006; Prandini et al., 2012) or feed-forward input (Calafiore and Fagiano, 2013b). While this computational approach allows for nearly arbitrary uncertainty in the system, the online optimization effort increases dramatically and recursive feasibility cannot be guaranteed. To alleviate the first problem, Zhang et al. (2014) as well as Schildbach et al. (2014) still use scenario theory and online sampling, but show how the number of samples can be significantly reduced.

Conceptual aspects, in particular rigorous closed-loop stability proofs and recursive feasibility, have received far less attention. Often only briefly addressed, a first significant exception was Primbs and Sung (2009) who highlighted the lack of recursive feasibility prevalent in many algorithms. Based on the idea of robust tube MPC (Langson et al., 2004; Mayne et al., 2005), a first rigorous solution of a stochastic constraint tightening was presented with the concept of "recursively feasible probabilistic tubes" (Kouvaritakis et al., 2010; Cannon et al., 2011), which was later extended to output feedback (Cannon et al., 2012). Therein, instead of considering the probability distribution ℓ steps ahead given the current state, the probability distribution ℓ steps ahead given any realization in the first $\ell - 1$ steps is considered. Alternatively, to obtain a larger admissible region, Korda et al. (2011) proposed a "first-step approach", where the constraints are replaced by a constraint which restricts the first predicted state to a robust invariant set. While recursive feasibility and satisfaction of the chance constraints could be proven, convergence of the closed loop could not be guaranteed. To prove recursive feasibility, Farina and Scattolini (2016) proposed to use the initial state of the prediction as an optimization variable. However, in this case the meaning of the probabilities remains unclear. An issue that has been addressed most recently by Hewing and Zeilinger (2018), where optionally using a previously predicted instead of the measured state as initial condition has been proposed and rigorously analysed.

As this review on stochastic MPC is by no means exhaustive but only touches some cornerstones and the most related publications, the interested reader is referred to the recent survey papers (Mayne, 2016; Mesbah, 2016; Farina et al., 2016) for further literature.

In summary, stochastic MPC, which is still a comparatively young topic, shows a similar development as nominal, stabilizing MPC. Initially, the main focus has been on tractable computational approaches, leaving a rigorous analysis of system theoretic conceptual aspects mostly aside. While successful practical applications and theoretical advances have been made, there are still many important open problems. Among them are (i) tractable algorithms that provide non-conservative recursive feasibility guarantees combined with a rigorous stability analysis, (ii) alleviation of the discussed shortcomings of sampling approaches which generally look promising, and (iii) stochastic stability and convergence results that are not based on a terminal constraint and terminal controller, which is in line with many practical implementations due to its simple, straight-forward design. After the development of the necessary background in Chapter 2, we address these topics in Chapter 3 and 5 of this thesis as detailed in Section 1.2.

MPC with model adaption Direct and indirect adaptive control has received considerable attention in the 1950s with significant progress and success. But enthusiasm declined sharply with disappointment and failures in practical application already in the 1960s. Yet, as emphasized in (Qin and Badgwell, 2003) as well as (Mayne, 2014), to reduce the cost of manual tuning in MPC and to cope with changing dynamics, there is still a strong interest in self-tuning predictive control formulations. However, being a notoriously difficult topic, only few rigorous solution strategies are available.

One of the first well-known *computational approaches* to adaptive receding horizon control was presented by Clarke et al. (1987a,b) and was "shown by simulation studies to be superior to accepted techniques". Yet, the approach does not discuss constraints and lacks a rigorous theoretical underpinning as argued in (Bitmead et al., 1990), where the authors attempt to correct this shortcoming. Genceli and Nikolaou (1996) introduced a persistence of excitation constraint in the MPC optimization to ensure parameter convergence without relying on an external excitation signal. Most notably, the authors derive a convex over-approximation with LMI constraints. Similarly, based on the fundamental results of Moore (1983), an input constraint which guarantees persistence of excitation is introduced in (Marafioti et al., 2014). A different approach, inspired by the notion of "dual control", which was presented in a series of seminal papers by Feldbaum (1961a,b), is taken in (Heirung et al., 2015, 2017), where an optimal solution is approximated by a Kalman filter for parameter estimation and an explicit cost term on the predicted covariance matrix.

General advanced identification methods for nonlinear systems have been successfully combined with linear robust MPC and nonlinear MPC, e.g., particle filtering (Bayard and Schumitzky, 2010). Most recently, Gaussian process regression has received an increasing attention due to its flexibility in modeling the uncertainty, e.g., (Klenske et al., 2016; Ostafew et al., 2015). Yet, efficiently propagating the state distribution remains a challenging problem and is subject to ongoing research (Hewing and Zeilinger, 2017).

Publications focusing on systems theoretic and algorithmic *conceptual aspects* include (Aswani et al., 2013) and (Di Cairano, 2016) where it is shown how a generic online system identification can be used within a robust MPC framework. In both contributions, recursive feasibility and constraint satisfaction is guaranteed by employing robust MPC methods based on an a priori given uncertainty set which is not updated online. System identification is used only to update a nominal prediction model in order to increase the closed-loop performance. Kim and Sugie (2008) introduced stabilizing adaptive MPC for linear SISO systems based on a modified least squares estimation but the approach and proofs crucially rely on the assumption of noise and additive disturbances being absent. Conceptual results for nonlinear systems have been presented in (Adetola and Guay, 2011; Guay et al., 2015), where set-membership system identification is proposed and stability properties are proven. The original algorithm leads to a computationally demanding, non-convex optimization program to be solved online. A computationally tractable combination of set-membership system identification and robust constraint tightening, yet restricted to uncertain FIR models, has been proposed by Tanaskovic et al. (2014) and successfully applied to a building control example (Tanaskovic et al., 2017). The algorithm has recently been extended by Bujarbaruah et al. (2018) with an additional linear Recursive Least Squares filter to handle chance constraints. In contrast to other set-membership techniques, in these contributions the uncertainty set is updated online and employed to reduce the conservatism in the predictions.

Summarizing, the competing goals of applicability with little system knowledge and satisfaction of constraints call for a trade-off, which is reflected in most publications. As the review above shows, current literature on the topic often either focuses on system identification and online learning in a practically applicable MPC algorithm or it provides a well defined, systems theoretic rigorous control algorithm but has limited applicability due to strong assumptions or intractable optimization algorithms to be solved. As detailed in the following section, it is the goal of this thesis to address this shortcoming and provide insights on predictive control for systems with model uncertainty. Specifically, in Chapter 4, we derive a computationally tractable, applicable approach that guarantees constraint satisfaction and stability for the closed-loop system.

1.2 Contributions and outline of the thesis

With this thesis, we contribute to the field of constrained control for systems with disturbances and uncertainties by addressing the above derived relevant open questions. In particular, we further the understanding of the conceptual aspects that are important for using stochastic and uncertain models within a moving horizon framework and derive rigorous computational approaches based thereon.

In the following, we outline in detail the contributions and the content of this thesis.

Chapter 2: Background In this chapter, we provide a brief review of nominal model predictive control, which is the foundation for the remainder of this thesis. Based thereon, a conceptual stochastic MPC framework is presented, discussed, and analyzed. In particular, we stress the important difference between the predicted states being modeled as random variables and the measurements revealing a particular realization thereof, which allows a local approximation of the stochastic optimal control program.

The results in the second part of this chapter have partially been presented in (Lorenzen et al., 2017d).

Chapter 3: Computational approaches for linear systems In this chapter, we take the step from the conceptual framework to computationally tractable algorithms. The objective is to understand sufficient properties of the online optimization to derive approximations and simplifications such that the desired closed-loop behavior proven for the conceptual algorithm still holds for the simplified stochastic MPC algorithm. The results presented in this chapter are applicable for linear systems subject to additive and multiplicative disturbances modeled by random variables that are pointwise in time independent and identically distributed. In order to derive the results, we separately discuss the requirements for feasibility of the optimization and stability of the closed loop. Thereby we highlight the difference between existence of a solution and feasibility of a suitable, a priori known candidate solution, which has not been exploited before.

The stochastic MPC algorithm presented in the first part of this chapter unifies previous results and leaves the designer the option to balance an increased feasible region against guaranteed bounds on the asymptotic average performance and convergence time. Besides the closed-loop properties of the conceptual algorithm given in Chapter 2, in addition we prove asymptotic stability in probability of the minimal robust positively invariant set that is stabilized by the unconstrained LQ optimal controller. In the second part of this chapter, motivated by results in statistical learning theory, finite sample approximations of the online stochastic program are proposed and discussed. Unlike earlier results based on scenario theory, we derive an MPC design based on samples drawn offline instead of independent samples in each iteration. To this end, we show that the constraint set can be separated into a state dependent nominal constraint set and a time-invariant chance constraint set, which is approximated offline. Thereby the design is reduced to a subset question for the chance constraint set and we provide a sufficient bound on the number of samples. For the closed-loop system, besides recursive feasibility and constraint satisfaction, asymptotic stability of the origin is proven under an additional assumption. The efficacy of the proposed approaches is demonstrated by numerical examples, which are used to underline and discuss the main theoretical properties.

The results of this chapter have previously been presented in (Lorenzen et al., 2015a,b, 2017b,c).

Chapter 4: Uncertainty and online model adaption In this chapter, we extend the problem setup to incorporate constant or slowly changing uncertainties, which amounts to output feedback as briefly discussed in Chapter 2. The objective is to simplify the conceptual algorithm in order to derive a theoretically rigorous, yet computationally tractable MPC algorithm, which retains the key property of online model adaption.

Similar to Chapter 3, the requirements for constraint satisfaction and closed-loop stability are considered separately. To prove recursive feasibility and robust constraint satisfaction, an online set-membership system identification is combined with homothetic prediction tubes. Additionally, an $\mathcal{H}_\infty$ optimal point estimate of the parameter is updated recursively to achieve a finite closed-loop gain from the disturbance to the state. Finally, we derive a persistence of excitation condition, under which convergence of the parameter estimates can be guaranteed. The proven properties and efficacy of the approach are illustrated in a numerical example at the end of the chapter.

While all results are derived for constant uncertainties, they directly extend to time-varying parameters. Furthermore, existing results in the MPC literature can be easily combined with the presented adaptive MPC algorithm, as it is given in a modern state space formulation. As a specific example, we present its application to a tracking task, where it leads to significantly improved results compared to a non-adaptive robust MPC.

The results of this chapter have previously been presented in (Lorenzen et al., 2017a, 2018).

Chapter 5: Stochastic MPC without terminal constraints In this chapter, we derive sufficient conditions for the stability of stochastic model predictive control without terminal cost and terminal constraints. The objective is to understand the basic system theoretic principles and derive applicable algorithms which remove the need of a terminal constraint, which is one of the main sources of conservatism.

Analogous to Chapter 2, we first study a conceptual MPC algorithm and provide stability results under the assumptions of optimization over general feedback laws and exact propagation of the probability density functions of the predicted states. We highlight why the stability proof, being based on the principle of optimality, does not directly extend to the computationally tractable approximations derived in Chapter 3. Based on these insights, stability results for tractable stochastic MPC algorithms using parameterized feedback laws and relaxation of the chance constraints are derived under stronger assumptions. The main results are presented for nonlinear systems along with examples and computational simplifications for linear systems.

The results of this chapter have previously been presented in (Lorenzen et al., 2017e,d).

Chapter 6: Conclusions In this chapter, we summarize the main results of this thesis and outline possible directions for future research.

Appendices Appendix A provides stability definitions of deterministic and stochastic systems that are employed throughout this thesis. Appendices B and C contain technical proofs and miscellaneous material that would otherwise distract from the main content and message.

Chapter 2

Background

As the results in this thesis are based upon fundamental results in stabilizing MPC, we briefly introduce a basic model predictive control algorithm for discrete-time systems in a mathematically rigorous way and present essential principles that are sufficient for closed-loop stability. Thereafter, in Section 2.2, a conceptual stochastic MPC algorithm is derived along with basic system theoretic properties of the closed loop. This serves as a basis for computational tractable approximations presented in the following chapters of this thesis.

2.1 A brief review of model predictive control

Consider a discrete-time dynamical system with state state $x_k \in \mathcal{X}$ and input $u_k \in \mathcal{U}$ that is modeled by the difference equation

$$x_{k+1} = f(x_k, u_k), \tag{2.1}$$

with given initial condition x_0, transition function $f : \mathcal{X} \times \mathcal{U} \to \mathcal{X}$, and time $k \in \mathbb{N}$. In light of Section 2.2, the state space $\mathcal{X}$ as well as the control value space $\mathcal{U}$ are assumed to be arbitrary metric spaces. The system is subject to hard constraints on the state and input

$$(x_k, u_k) \in \mathbb{Z}, \tag{2.2}$$

which should be satisfied point-wise in time for all $k \in \mathbb{N}$ and a given constraint set $\mathbb{Z} \subseteq \mathcal{X} \times \mathcal{U}$. The control objective is to find a feedback law which stabilizes the system (2.1) at a given setpoint $\bar{x}$ such that the constraints (2.2) are satisfied for all $k \in \mathbb{N}$. In the following, the setpoint $\bar{x}$ is assumed to be a feasible equilibrium of system (2.2), that is, there exists a control input $\bar{u}$ such that $\bar{x} = f(\bar{x}, \bar{u})$ and $(\bar{x}, \bar{u}) \in \mathbb{Z}$. The performance of the control system is measured by a running cost $\ell : \mathcal{X} \times \mathcal{U} \to \mathbb{R}_{\geq 0}$, which penalizes the distance to the desired setpoint and the control input.

To solve this control task, the basic idea of MPC is to approximate the optimal control problem locally by repeatedly solving a finite horizon optimal control problem[1]

$$V_N(x_k) = \min_{\mathbf{u}_{N|k}} \sum_{l=0}^{N-1} \ell(x_{l|k}, u_{l|k}) + V_f(x_{N|k}) \tag{2.3a}$$

$$\text{s.t. } x_{l+1|k} = f(x_{l|k}, u_{l|k}), \qquad x_{0|k} = x_k \tag{2.3b}$$

$$(x_{l|k}, u_{l|k}) \in \mathbb{Z}, \qquad \forall l \in \mathbb{N}_0^{N-1} \tag{2.3c}$$

$$x_{N|k} \in \mathbb{X}_f, \tag{2.3d}$$

where the predicted input sequence is denoted by $\mathbf{u}_{N|k} = (u_{0|k}, \cdots, u_{N-1|k})$ and similarly the corresponding state sequence by $\mathbf{x}_{N|k} = (x_{1|k}, \cdots, x_{N|k})$. The initial condition $x_{0|k} = x_k$ is the measured state at time k and $N \geq 2$ the prediction horizon. To guarantee closed-loop stability, the terminal cost $V_f : \mathcal{X} \to \mathbb{R}_{\geq 0}$ as well as the terminal constraint $\mathbb{X}_f \subseteq \mathcal{X}$ need to be suitably chosen, as discussed below. Throughout this thesis, the double index notation is used to distinguish predicted states and inputs from realized states and inputs, e.g., $u_{l|k}$ for the input predicted l steps ahead at time k and u_{l+k} for the input realized at time $l+k$. Assuming a minimizer $\mathbf{u}^*_{N|k}$ exists, the MPC algorithm can be summarized as follows.

Algorithm 2.1 (Basic MPC algorithm).
Offline: Given the constraint set $\mathbb{Z}$ and the running cost ℓ, determine a suitable terminal cost V_f, terminal constraint $\mathbb{X}_f$, and prediction horizon N.
Online: At each time step $k = 0, 1, 2, \ldots$

1. Measure the current state x_k.
2. Determine a minimizer $\mathbf{u}^*_{N|k}$ of the optimal control problem (2.3).
3. Apply the MPC feedback law $\kappa(x_k) = u^*_{0|k}$.

One of the main conceptual questions in MPC are concerned with finding sufficient conditions and constructive design approaches which ensure existence of an optimal solution at each sampling instant, thus guaranteeing a well defined control law and asymptotic stability of the desired steady state in closed loop. The following standard assumptions and results for the basic MPC algorithm are well known and can be found in, e.g., (Mayne et al., 2000; Rawlings et al., 2017; Grüne and Pannek, 2017).

[1] In line with most literature on MPC, throughout this thesis, $\mathbf{x}_{N|k}$ is omitted as an argument of the minimization operator since it is explicitly defined by the transition function f.

Let the set of feasible decision variables of the optimization (2.3) for a given state x_k be defined as

$$\mathbb{D}_N(x_k) = \{\mathbf{u}_{N|k} \in \mathcal{U}^N \mid \exists \mathbf{x}_{N|k} \text{ satisfying } (2.3\text{b}) - (2.3\text{d})\}$$

and denote the set of feasible initial conditions by $\mathbb{X}_N = \{x \in \mathcal{X} \mid \mathbb{D}_N(x) \neq \emptyset\}$.

Assumption 2.1 (Continuity). *The transition function f, running cost ℓ, and terminal cost V_f are continuous.*

Assumption 2.2 (Compactness). *For each $x \in \mathcal{X}$ the set $\mathbb{D}_N(x)$ is compact.*

Assumption 2.3 (Basic stability assumption). *The terminal constraint set $\mathbb{X}_f \subseteq \mathbb{X}$ is closed and contains the setpoint $\bar{x}$ in its interior. There exists a local control law $\kappa_f : \mathbb{X}_f \to \mathcal{U}$ such that the following is satisfied for all $x \in \mathbb{X}_f$:*

i) $(x, \kappa_f(x)) \in \mathbb{Z}$

ii) $f(x, \kappa_f(x)) \in \mathbb{X}_f$

iii) $V_f(f(x, \kappa_f(x))) + \ell(x, \kappa_f(x)) \leq V_f(x)$

There exist $\mathcal{K}_\infty$ functions α_1, α_f satisfying

$$\begin{aligned} &\alpha_1(\|x\|_{\bar{x}}) \leq \ell(x, u) && \forall x, u \text{ with } x \in \mathbb{X}_N, (x, u) \in \mathbb{Z} \\ &V_f(x) \leq \alpha_f(x) && \forall x \in \mathbb{X}_f \end{aligned}$$

Assumption 2.1 and 2.2 imply, by the extreme value theorem, existence of an optimal solution to problem (2.3) for $x_k \in \mathbb{X}_N$. Assumption 2.2 is satisfied for example if $\mathbb{Z}$ is compact, which is often assumed for simplicity. It could be further relaxed if the cost function is coercive (Rawlings et al., 2017). Assumption 2.3 implies that the terminal set $\mathbb{X}_f$ is a control invariant set for the constrained system and hence any state sequence ending in $\mathbb{X}_f$ can be suitably extended, remaining in $\mathbb{X}_f$. Furthermore, V_f serves as a local control Lyapunov function which allows to establish that the optimal value function V_N is a valid Lyapunov function for the closed loop. The lower bound on the running cost can be relaxed under an observability condition and the assumption of $\mathbb{X}_f$ having an interior can be relaxed under a controllability assumption. Constructive approaches on how to obtain suitable choices for $\mathbb{X}_f$ and V_f based on a local linearization can be found in (Rawlings et al., 2017), going back to (Michalska and Mayne, 1993; Chen and Allgöwer, 1998).

Under the above assumptions, it can be shown that the basic MPC algorithm is a well defined control law for the nominal system (2.1) which asymptotically stabilizes the desired set-point.

Lemma 2.2 (Existence of a solution). *Suppose Assumptions 2.1 and 2.2 are satisfied. If $x_k \in \mathbb{X}_N$, then problem (2.3) is feasible and an optimal solution $\mathbf{u}^*_{N|k}$ exists.*

Theorem 2.3 (MPC closed-loop properties). *Consider system 2.1 in closed loop with the basic MPC Algorithm 2.1 and suppose Assumptions 2.1-2.3 are satisfied. Optimization (2.3) is recursively feasible, that is $x_k \in \mathbb{X}_N$ implies $f(x_k, \kappa(x_k)) \in \mathbb{X}_N$. Furthermore, x^* is an asymptotically stable equilibrium point with region of attraction $\mathbb{X}_N$ for the closed-loop system $x_{k+1} = f(x_k, \kappa(x_k))$ and the constraints (2.2) are satisfied by the state and input trajectories if $x_0 \in \mathbb{X}_N$.*

A proof of Theorem 2.3 can be found in (Grüne and Pannek, 2017) or similarly under the restriction $\mathcal{X} = \mathbb{R}^n$ and $\mathcal{U} = \mathbb{R}^m$ in (Mayne et al., 2000; Rawlings et al., 2017). The key idea is to first show that, under Assumption 2.3, the optimal value function is a valid Lyapunov function candidate and then exploit the auxiliary control law κ_f to construct a feasible candidate solution which satisfies a cost decrease and implies recursive feasibility. This basic idea is a recurring theme in proving stability of different MPC algorithms, including some presented in this thesis.

While concepts based on Assumption 2.3 are most common, conceptually different sufficient conditions and stability proofs exists. In particular, results based on a controllability assumption and sufficiently long prediction horizon instead of a terminal constraint and cost as presented in, e.g., (Alamir and Bornard, 1995; Grimm et al., 2005; Grüne and Rantzer, 2008). Although being of particular interest for systems under uncertainty, the stability proofs are based on the principle of optimality and therefore, as presented in Chapter 5, do not trivially extend to approximations used in robust and stochastic MPC.

In summary, for deterministic systems there exists a solid theoretical foundation for solving the constrained set-point stabilization problem within an MPC framework. In particular, sufficient conditions for stability have been distilled along with constructive approaches to design the optimal control problem.

2.2 Stochastic MPC – A conceptual algorithm

In this section, we introduce stochastic model predictive control on a conceptual level, thereby providing the theoretical foundation that leads to the practically feasible algorithms presented in the following chapters. While the conceptual ideas are implicit in most stochastic MPC publications, they are, with an exception being (Sehr and Bitmead, 2018), not made explicit in the available literature. In particular, the explicit distinction of predictions modeled as random variables and the measured state of the system being a realization in $\mathbb{R}^n$, which allows to derive suitable local approximations, has, to the best of our knowledge, not been discussed. The following results have partially been presented in (Lorenzen et al., 2017d).

Consider a stochastic discrete time system given by

$$x_{k+1} = f(x_k, u_k, w_k) \tag{2.4}$$

with state $x_k \in \mathbb{R}^n$, controlled input $u_k \in \mathbb{R}^m$, and exogenous disturbance input $w_k \in \mathbb{R}^{m_w}$ on which we take the following assumption.

Assumption 2.4 (System and disturbance). *The system dynamics are given by a continuous function f with $f(0,0,0) = 0$. The disturbances w_k, $k \in \mathbb{N}$ are realizations of independent and identically distributed, zero-mean random variables $W_k : \Omega \to \mathbb{W}$ defined on the probability space $(\Omega, \mathcal{B}, \mathbb{P}_w)$ with $\mathbb{W} \subseteq \mathbb{R}^{m_w}$ and probability density function ρ_w.*

In the following, consider the product probability $\mathbb{P} = \otimes_{k=0}^{\infty} \mathbb{P}_w$, denote by $\mathbb{P}_k\{A\} = \mathbb{P}\{A \mid x_k\}$ the probability of an event A given the realized state at time k, and similarly the conditional expected value $\mathbb{E}_k\{\cdot\} = \mathbb{E}\{\cdot \mid x_k\}$. Furthermore, the abbreviation A $\mathbb{P}$ *a.s.* is used for $\mathbb{P}\{A\} = 1$, i.e., the probability of the event A occurring is one.

With the state x_k at time k, predicted, future states are given by

$$x_{l+1|k} = f(x_{l|k}, u_{l|k}(x_{l|k}), W_{l+k}), \qquad x_{0|k} \overset{a.s.}{=} x_k \tag{2.5}$$

where $x_{l|k}$ are random variables[2] and $u_{l|k} : \mathbb{R}^n \to \mathbb{R}^m$ are Borel measurable functions.

We explicitly highlight the difference between (2.4) and (2.5) as a sharp distinction has to be drawn between predictive control of a process where the state is modeled by a *distribution* and a process with state in $\mathbb{R}^n$ that is modeled as the *realization* of a stochastic process. For example, if an ensemble modeled as distribution is to be controlled, the measurements can be aggregated data or a vast amount of samples that, idealized, can be considered as *measuring the distribution* itself. If, on the other hand, a probability distribution is used to model disturbances or uncertainties, measurement data can be viewed as a realization of the probability distribution, i.e., there exists *one realization*, which is revealed by *one* measurement. In the former case, the state x_k is an element of a function space, which is covered by Section 2.1. In the latter case, we are interested in a specific realization, a state $x_k \in \mathbb{R}^n$, which can be estimated or measured online and only the predictions are random variables. While the first is more general and subsumes

[2]The convention of using capital letters for random variables and lower case letters for realizations is not respected here in order to comply with common MPC notation.

the second as a special case, for the latter computationally much more tractable locally approximating formulations can be found and significant progress has been made in recent years.

As in the previous section, the controlled system is subject to hard constraints on the input

$$u_{l|k}(x_{l|k}) \in \mathbb{U} \quad \mathbb{P}_k \ a.s. \tag{2.6}$$

with $\mathbb{U} = \{u \in \mathbb{R}^m \mid c_u(u) \leq 0\}$, $c_u : \mathbb{R}^m \to \mathbb{R}$. Additionally, the stochastic model allows to incorporate probabilistic constraints on the state

$$\mathbb{P}\{x_{l+1|k} \in \mathbb{X} \mid x_{l|k}\} \geq 1 - \varepsilon \quad \mathbb{P}_k \ a.s. \tag{2.7}$$

with $\mathbb{X} = \{x \in \mathbb{R}^n \mid c_x(x) \leq 0\}$, $c_x : \mathbb{R}^n \to \mathbb{R}$ and $\varepsilon \in [0, 1]$. Note that the definition of the chance constraint (2.7) is in line with a receding horizon control law as it requires that the state $x_{l+1|k}$ does not violate the constraints with probability at least $1 - \varepsilon$ independent of the realization of the previous $x_{l|k}$.

The performance of the control system is measured by the expected value of a running cost $\ell : \mathbb{R}^n \times \mathbb{R}^m \to \mathbb{R}_{\geq 0}$, which as before penalizes the distance to the desired setpoint $\bar{x} = 0$ and input $\bar{u} = 0$. With a suitably chosen terminal cost $V_f : \mathbb{R}^n \to \mathbb{R}_{\geq 0}$ and terminal constraint set $\mathbb{X}_f = \{x \in \mathbb{R}^n \mid c_f(x) \leq 0\}$, $c_f : \mathbb{R}^n \to \mathbb{R}$, the stochastic optimal control problem, which is to be solved at each time step k, is given by

$$V_N(x_k) = \min_{\mathbf{u}_{N|k}} \mathbb{E}_k \left\{ \sum_{l=0}^{N-1} \ell(x_{l|k}, u_{l|k}(x_{l|k})) + V_f(x_{N|k}) \right\} \tag{2.8a}$$

$$\text{s.t. } x_{l+1|k} = f(x_{l|k}, u_{l|k}(x_{l|k}), W_{l+k}), \qquad x_{0|k} \stackrel{a.s.}{=} x_k \tag{2.8b}$$

$$\mathbb{P}\{x_{l+1|k} \in \mathbb{X} \mid x_{l|k}\} \geq 1 - \varepsilon \quad \mathbb{P}_k \ a.s. \quad \forall l \in \mathbb{N}_0^N \tag{2.8c}$$

$$u_{l|k}(x_{l|k}) \in \mathbb{U} \quad \mathbb{P}_k \ a.s. \qquad \forall l \in \mathbb{N}_0^{N-1} \tag{2.8d}$$

$$x_{N|k} \in \mathbb{X}_f \quad \mathbb{P}_k \ a.s. \tag{2.8e}$$

As before, with $\mathcal{U}$ being the set of all measurable functions $u : \mathbb{R}^n \to \mathbb{R}^m$, we denote the set of feasible decision variables of the SMPC optimization (2.8) for a given state x_k by

$$\mathbb{D}_N(x_k) = \{\mathbf{u}_{N|k} \in \mathcal{U}^N \mid \exists \mathbf{x}_{N|k} \text{ satisfying (2.8b) – (2.8e)}\}$$

and the set of feasible initial conditions by $\mathbb{X}_N = \{x \in \mathbb{R}^n \mid \mathbb{D}_N(x) \neq \emptyset\}$. To assess the properties of the optimal control problem and the closed-loop system we introduce the following assumptions.

Assumption 2.5 (Continuity). *The stage cost ℓ, terminal cost V_f, as well as constraint functions c_u, c_x, and c_f are continuous.*

Assumption 2.6 (Basic stability assumption, stochastic MPC). *There exists a non-empty terminal constraint set $\mathbb{X}_f$, a local control law $\kappa_f : \mathbb{X}_f \to \mathbb{R}^m$, and a non-negative constant $c_1 \in \mathbb{R}_{\geq 0}$ such that the following is satisfied for all $x \in \mathbb{X}_f$:*

i) $\mathbb{P}_w\{f(x, \kappa_f(x), W) \in \mathbb{X}\} \geq 1 - \varepsilon$, $\kappa_f(x) \in \mathbb{U}$

ii) $f(x, \kappa_f(x), W) \in \mathbb{X}_f \quad \mathbb{P}_w$ *a.s.*

iii) $\mathbb{E}_w\{V_f(f(x, \kappa_f(x), W))\} + \ell(x, \kappa_f(x)) \leq V_f(x) + c_1$.

There exists a continuous, positive definite, radially unbounded function $\sigma : \mathbb{R}^n \to \mathbb{R}_{\geq 0}$ such that $\ell(x, u) \geq \sigma(x)$ for all $u \in \mathbb{U}$.

Assumption 2.7 (Controllability). *There exists a scalar $\bar{a} > 1$ such that $V_N(x) \leq \bar{a}\sigma(x) + V_N(0)$.*

Continuity of the cost and constraint function implies that the expected value and probabilities are well defined (Klenke, 2014, Theorem 1.88). Assumption 2.6 slightly differs from Assumption 2.3, in particular through the additional constant c_1. This is necessary because in the case of a persistent additive disturbance, the origin cannot be a steady state of the closed-loop system. In this setup, it would hence be natural to consider the evolution of the distribution of all possible realizations (2.4) and a stage cost which is positive definite with respect to a desired steady-state distribution and optimal control law. This leads to the well established theory presented in the previous section. Yet, as discussed above, we do not proceed this way for two main reasons: First, defining a desired steady-state distribution and control law is, from an application point of view, in general quite difficult. Second, we emphasize again that the goal is not to control a probability distribution but a single realization of a stochastic process for which tractable, local approximations are discussed in the following chapters. Finally, similar to nominal MPC, under Assumption 2.7 stronger stability notions can be proven.

Assuming a minimizer $\mathbf{u}^*_{N|k}$ exists, the stochastic MPC algorithm is equivalent to Algorithm 2.1 and the SMPC control law is defined by $\kappa(x_k) = u^*_{0|k}(x_k)$. The closed-loop properties can be summarized as follows.

Proposition 2.4 (SMPC closed-loop properties). *Consider the stochastic MPC algorithm based on optimization* (2.8). *Suppose Assumptions 2.4-2.6 are satisfied, $x_0 \in \mathbb{X}_N$, and a minimizer $\mathbf{u}^*_{N|k} \in \mathcal{U}$ exists for all $x_k \in \mathbb{X}_N$.*

The stochastic MPC feedback law $\kappa(x_k) = u^*_{0|k}(x_k)$ applied to system 2.4 renders the set $\mathbb{X}_N$ forward invariant, that is $x_k \in \mathbb{X}_N$ implies $f(x_k, \kappa(x_k), W_k) \in \mathbb{X}_N$ $\mathbb{P}_w$ a.s. For all

$k \in \mathbb{N}$, the closed-loop trajectories satisfy the input constraints $u_k \in \mathbb{U}$ almost surely and for each $k \in \mathbb{N}$ the state constraints $x_k \in \mathbb{X}$ with at least probability $1-\varepsilon$. Furthermore it holds

$$\mathbb{E}_k\left\{V_N(x_{k+1})\right\} - V_N(x_k) \leq -\ell(x_k, u_k) + c_1. \tag{2.9}$$

If additionally Assumption 2.7 is satisfied, then there exists a scalar $\lambda \in (0,1)$ such that

$$\mathbb{E}_k\left\{V_N^0(x_{k+1})\right\} \leq \lambda V_N^0(x_k) + c_1 \tag{2.10}$$

where $V_N^0(x) = V_N(x) - V_N(0)$.

The proof of Proposition 2.4 does not reveal relevant insight and can hence be found in Appendix B.1 for the sake of completeness. By standard results, Proposition 2.4 directly implies different stochastic stability notions of the closed-loop system, in particular properties of the limiting behavior of x_k and bounds on the probability that sample paths of the state trajectory remain within level sets of the optimal value function.

Corollary 2.5. *Suppose Assumptions 2.4-2.6 are satisfied. For the closed-loop system $x_{k+1} = f(x_k, \kappa(x_k), W_k)$ with $x_0 \in \mathbb{X}_\infty$ it holds*

$$\lim_{n\to\infty} \frac{1}{n} \sum_{k=0}^{n-1} \mathbb{E}_0\left\{\sigma(x_k)\right\} \leq c_1.$$

If $\lim_{k\to\infty} \mathbb{E}_0\{\sigma(x_k)\}$ exists, then it is at most c_1.

Corollary 2.6. *Suppose Assumptions 2.4-2.7 are satisfied. For sample paths of the closed-loop system $x_{k+1} = f(x_k, \kappa(x_k), W_k)$ with $x_0 \in \mathbb{X}_\infty$ it holds*

$$\mathbb{P}_0\{\sup_{K\geq k\geq 0} V_N(x_k) \geq m\} \leq V_N(x_0)\frac{\lambda^K}{m} + \frac{1-\lambda^K}{1-\lambda}\frac{c_1}{m}.$$

Corollary 2.7. *Suppose Assumptions 2.4-2.7 are satisfied with $c_1 = 0$. For the closed-loop system $x_{k+1} = f(x_k, \kappa(x_k), W_k)$, the origin is exponentially asymptotically stable with probability 1 and the region of attraction $\mathbb{X}_\infty$. If σ can be chosen as $\sigma(x) = a_l\|x\|_2^2$, then the origin is asymptotically mean square stable.*

In general, when $c_1 \neq 0$ and $\sigma(x)$ is large, the distance of the state to the origin decreases "on average" along trajectories of the process and ultimately "oscillates" in some random fashion around the origin. Corollary 2.5 provides, similar to an ultimate bound for deterministic systems, a bound on the asymptotic average distance of the state to the origin. In contrast, Corollary 2.6 provides an explicit bound on the probability that a sample path leaves a region of interest within a certain time frame. Finally, in Corollary 2.7 an asymptotic stability statement is made analogous to the results in the previous section. Given Proposition 2.4, Corollary 2.5 follows by taking iterated expectations, Corollary 2.6 and 2.7 follow by standard results in martingale theory and stochastic control, cf. (Kushner, 1967, Theorem 3.III), (Kushner, 2014).

In the following, we briefly discuss the optimization program (2.8) and different formulations thereof, which are important in the next chapters and for computational approaches to SMPC in general.

As comprehensively discussed in (Bertsekas and Shreve, 1978, Chapter 3), the SMPC optimization can equivalently be formulated using the stochastic dynamic programming principle. In particular, for $N > 1$ we have

$$
\begin{aligned}
V_N(x_k) = \min_{u_k \in \mathbb{R}^m} \ & \left\{ \ell(x_k, u_k) + \int_{\mathbb{W}} V_{N-1}(f(x_k, u_k, w)) \mathbb{P}_w(\mathrm{d}w) \right\} \\
\text{s.t. } \ & u_k \in \mathbb{U} \\
& \mathbb{P}_w\{f(x_k, u_k, W) \in \mathbb{X}\} \geq 1 - \varepsilon \\
& f(x_k, u_k, W) \in \mathbb{X}_{N-1} \quad \mathbb{P}_w \ a.s.
\end{aligned}
\tag{2.11}
$$

Based thereon, computational approaches to stochastic MPC have been developed, e.g., (Batina, 2004). However, the proposed algorithms generally suffer from a high computational complexity and the "curse of dimensionality" if gridding the state space is necessary. In the following chapters, the reformulation in terms of equation (2.11) will not be exploited for computational algorithms but still be used in the stability proofs, in particular in Chapter 5, where stability without terminal constraints is discussed.

Using the transition kernel of the controlled Markov process $x_{l|k}$, another equivalent formulation of optimization (2.8) can be given directly in terms of the evolution of the probability density functions ρ_l of the predicted states $x_{l|k}$. Let $\kappa_u : \mathcal{B} \times \mathbb{R}^n \to [0, \infty)$ be the stochastic state transition kernel defined by $\kappa_u(A|x) = \mathbb{P}_w \{w \text{ s.t. } f(x, u, w) \in A | x, u\}$ for all $A \in \mathcal{B}$. Under the given assumptions, κ_u admits a density ρ_u such that $\kappa_u(A|x) = \int_A \rho_u(y|x) \mathrm{d}y$. Optimization (2.8) can then equivalently be written as

$$\begin{aligned}
\min_{\mathbf{u}_{N|k}} \; & \sum_{l=0}^{N-1} \int_{\mathbb{R}^n} \ell(x, u_{l|k}(x))\rho_{l|k}(x)\mathrm{d}x + \int_{\mathbb{R}^n} V_f(x)\rho_{N|k}(x)\mathrm{d}x \\
\text{s.t. } & \rho_{l+1|k}(x) = \int_{\mathbb{R}^n} \rho_{u_{l|k}}(x|y)\rho_{l|k}(y)\mathrm{d}y, \rho_{0|k}(x) = \delta(x - x_k) \\
& \int_{\mathbb{R}^n} \mathbb{1}_{[1-\varepsilon,1]}(\kappa_{u_{l|k}}(\mathbb{X}|x))\rho_{l|k}(x)\mathrm{d}x = 1 \qquad \forall l \in \mathbb{N}_0^{N-1} \\
& \int_{\mathbb{R}^n} \mathbb{1}_{\mathbb{U}}(u_{l|k}(x))\rho_{l|k}(x)\mathrm{d}x = 1 \qquad \forall l \in \mathbb{N}_0^{N-1} \\
& \int_{\mathbb{R}^n} \mathbb{1}_{\mathbb{X}_f}(x)\rho_{N|k}(x)\mathrm{d}x = 1
\end{aligned} \tag{2.12}$$

where δ denotes the Dirac delta function and $\mathbb{1}_A(\cdot)$ the indicator function on a set A. Note that alternatively, if $f(\cdot, u(\cdot), w)$ is bijective and continuously differentiable, then the state probability density function can as well be given explicitly by a change of variables and marginalisation (Klenke, 2014, Theorem 1.101)

$$\rho_{l+1|k}(x) = \int_{\mathbb{R}^{m_w}} \frac{\rho_{l|k}(f^{-1}(x, u_{l|k}(x), w))}{|\det(J_f(f^{-1}(x, u_{l|k}(x), w)))|}\rho_w(w)\mathrm{d}w$$

where J_f denotes the Jacobian of $f(\cdot, u(\cdot), w)$, f^{-1} its inverse, and ρ_w the probability density function of W_k. Similarly, as in nominal MPC, a "condensed form" based on explicitly iterating the transition function can formulated.

This formulation in terms of probability density functions, while still infinite dimensional, gives rise to finite dimensional numerical approximations based on the use of basis functions for the control functions $u_{l|k}$ and numerical integration schemes. To overcome the curse of dimensionality, in the following chapter, we introduce a randomized sampling approach to approximate the integrals and explicitly consider the integration error made with finite sample sets.

Remark 2.8. The results presented in this section apply similarly if the state is not fully measurable but only an output $y_k = h(x_k, v_k)$ with iid measurement noise $V_k \sim \mathbb{P}_v$ is assumed. In this case, the initial condition in (2.5) is not almost surely determined by a point in $\mathbb{R}^n$ but by a random variable $x_{0|k}$ whose conditional probability density function $\rho_{0|k}$ given the past input and output data can be derived by a Bayesian filter as

$$\rho_{0|k}(x) = \frac{\rho_{y_k}(x)\rho_{1|k-1}(x)}{\int_{\mathbb{R}^n} \rho_{y_k}(x)\rho_{1|k-1}(x)\mathrm{d}x} \tag{2.13}$$

with $\rho_{y_k}(x)$ being the probability density function describing the conditional distribution of observing the output y given the state x. Similarly, the input is not a function mapping points in $\mathbb{R}^n$ to the input space $\mathbb{R}^m$, but rather one mapping the probability density function of $x_{0|k}$ to $\mathbb{R}^m$. We have not presented stochastic MPC in this generality, as the clear interpretation of using MPC to derive a local approximation to an infinite horizon optimal control problem would have been lost. Additionally, there are only few cases, most notably the one of linear systems with additive Gaussian disturbance, where the posterior distribution of the state, i.e. $x_{0|k}$, has a computationally tractable, finite dimensional representation. However, one example of particular interest that can be modeled in this output feedback framework is stochastic MPC for dynamical systems with parametric *uncertainty*. This is discussed in Chapter 4, albeit, due to the mentioned lack of a finite dimensional representation, in a deterministic framework.

Remark 2.9. We highlight the relation of the presented stochastic MPC to economic MPC, where a more general stage cost is considered, which is not necessarily positive definite with respect to a desired steady state. In particular, in (Rawlings et al., 2008) a positive definite stage cost with respect to an unreachable steady state has been studied in a deterministic setup. The authors report a better closed-loop performance in the transient compared to using a cost which is positive definite with respect to the best reachable steady state. It remains an open question whether the improved performance translates to stochastic MPC when comparing a cost which is positive definite with respect to a steady-state distribuition versus, as considered here, a cost which is positive definite with respect to a desired set-point.

2.3 Summary

In this chapter, we rigorously presented a basic stabilizing nominal MPC as well as a conceptual stochastic MPC algorithm and recalled relevant system theoretic properties of the resulting closed loop. In particular, the important difference between the predicted states being random variables and the measurements revealing a particular realization thereof has been discussed. This builds the foundation for the computational approaches presented in the following chapters.

Chapter 3

Computational approaches to stochastic MPC for linear systems

In this chapter, we introduce computational approaches to stochastic MPC for the special class of linear systems with additive and multiplicative disturbances modeled by a stochastic process. In particular, rigorous design methods are derived based on the separation principle and randomized sampling methods to approximate the optimal control problem presented in the previous chapter. In contrast to generic numerical integration schemes for solving (2.12), the proposed method does not suffer from the curse of dimensionality. In the design and analysis we explicitly take into account the advantages and shortcomings of approximate methods. In particular, the chance constraints $\mathbb{P}\{x_{l+1|k} \in \mathbb{X} \mid x_{l|k}\} \geq 1-\varepsilon$ are relaxed to $\mathbb{P}\{x_{l+1|k} \in \mathbb{X} \mid x_k\} \geq 1-\varepsilon$ and its implications on recursive feasibility and stability are discussed. This relaxation decreases conservatism but results in a lack of recursive feasibility, which is subsequently addressed by an additional "first step constraint".

In the first part of this chapter, linear systems with *additive disturbances* are addressed. We exploit linearity to separate the predicted states into a nominal state sequence, which is deterministic, and an error sequence, which is a martingale. The latter is used to design a constraint set for the former such that the stochastic optimal control program is reduced to a deterministic, linearly constrained quadratic program. The introduced stochastic MPC algorithm is novel and unifies previous results, in particular the use of "recursively feasible tubes" (Kouvaritakis et al., 2010) and the "least restrictive approach" (Korda et al., 2011).

In the second part, linear systems with *multiplicative disturbances* are addressed. In this setup, the superposition principle does not apply and a simple "constraint tightening" approach cannot be used. We reformulate the problem to exploit results from statistical learning theory and provide precise statements on the sample complexity such that, with a user-specified confidence, constraint satisfaction is guaranteed. In particular, we reduce the question of constraint satisfaction to the question whether a certain random set, derived by sampling the constraints, is a subset of a deterministic set given through probabilistic constraints.

The results presented in this chapter are based on (Lorenzen et al., 2015a,b, 2017b,c).

3.1 Linear systems with additive disturbances

3.1.1 Problem setup and surrogate model

In this section, we assume that the system dynamics (2.4) are linear and time-invariant, i.e.,

$$x_{k+1} = Ax_k + Bu_k + B_w w_k. \tag{3.1}$$

As before, the disturbance sequence $(w_k)_{k\in\mathbb{N}}$ is assumed to be a realization of a stochastic process $(W_k)_{k\in\mathbb{N}}$ satisfying the following assumption.

Assumption 3.1 (Disturbance sequence). *W_k for $k \in \mathbb{N}$ are independent and identically distributed, zero mean random variables with distribution $\mathbb{P}_w$ and support $\mathbb{W} \subset \mathbb{R}^{m_w}$. The set $\mathbb{W}$ is bounded and convex.*

In contrast to Assumption 2.4, the set $\mathbb{W}$ is assumed to be bounded in order to derive explicit guarantees for recursive feasibility. A relaxation to a bounded confidence region is briefly touched upon in Section 3.1.5. The zero-mean assumption could be relaxed without changing the following results by adding the mean value as additional deterministic input.

As introduced in Chapter 2, we denote the infinite product probability by $\mathbb{P} = \otimes_{k=0}^{\infty}\mathbb{P}_w$, the probability of an event A given the realized state at time k by $\mathbb{P}_k\{A\} = \mathbb{P}\{A \mid x_k\}$, and similarly the expected value by $\mathbb{E}_k\{\cdot\} = \mathbb{E}\{\cdot \mid x_k\}$. Given the state x_k at time k, the predicted, future states are modeled by

$$x_{l+1|k} = Ax_{l|k} + Bu_{l|k}(x_{l|k}) + B_w W_{l+k}, \qquad x_{0|k} \overset{a.s.}{=} x_k$$

where $x_{l|k}$ are random variables and $u_{l|k} : \mathbb{R}^n \to \mathbb{R}^m$ are measurable functions in $x_{l|k}$. The state and input constraint sets are assumed to be given by convex polytopes $\mathbb{X} = \{x \in \mathbb{R}^n \mid Hx \leq h\}$ and $\mathbb{U} = \{u \in \mathbb{R}^m \mid Gu \leq g\}$ with $H \in \mathbb{R}^{p\times n}$, $G \in \mathbb{R}^{q\times m}$, $h \in \mathbb{R}^p$, $g \in \mathbb{R}^q$. Yet, due to computational reasons discussed in Section 3.1.5, we assume individual chance constraints for each hyperplane, that is constraints

$$\mathbb{P}_w\{[H]_j x_{1|k} \leq [h]_j \mid x_k\} \geq 1 - [\varepsilon]_j \qquad \forall j \in \mathbb{N}_1^p, \tag{3.2a}$$

$$Gu_k \leq g \tag{3.2b}$$

with $\varepsilon \in [0,1]^p$. Inequality (3.2a) restricts to $[\varepsilon]_j$ the probability of violating the linear state constraint given by the j-th hyperplane at the next sampling time, independent of the realization of the state x_k at time k.

Finally, the running cost ℓ is assumed to be given by the quadratic function $\ell(x,u) = \|x\|_Q^2 + \|u\|_R^2$ with $Q \in \mathbb{R}^{n\times n}$, $Q \succ 0$, $R \in \mathbb{R}^{m\times m}$, $R \succ 0$.

Nominal system and optimization program

In order to derive a practically feasible approach, we need to sacrifice optimality for computational simplicity. In the following, we introduce an input parameterization, a nominal system, and simplified constraints that will be designed later such that their satisfaction is sufficient for adherence of the original constraints (3.2) by the true system.

Let the state of the nominal system be denoted by $z_{l|k}$ and be defined through the expected value of the predicted states, that is $z_{l|k} = \mathbb{E}_k\{x_{l|k}\}$. Then the error sequence $(e_{l|k})_{l\in\mathbb{N}} = (x_{l|k} - z_{l|k})_{l\in\mathbb{N}}$, defined by the deviation of the predicted states from the nominal states, is a martingale and in particular $\mathbb{E}\{e_{l|k}\} = 0$. The input functions $u_{l|k}(\cdot)$ are parameterized[1] by

$$u_{l|k}(x_{l|k}) = Ke_{l|k} + v_{l|k}$$

with free optimization variables $v_{l|k} \in \mathbb{R}^m$ and constant prestabilizing feedback gain $K \in \mathbb{R}^{m\times n}$ such that $A_{cl} = A + BK$ is Schur stable. Summarizing, the dynamics of the nominal and error system are given by

$$z_{l+1|k} = Az_{l|k} + Bv_{l|k}, \qquad z_{0|k} = x_k, \tag{3.3a}$$

$$e_{l+1|k} = A_{cl}e_{l|k} + B_w W_{l+k}, \qquad e_{0|k} \overset{a.s.}{=} 0. \tag{3.3b}$$

With $\ell(x,u) = \|x\|_Q^2 + \|u\|_R^2$, the terminal cost is defined by $V_f(x) = \|x\|_P^2$, where P is the solution to the discrete-time Lyapunov equation $A_{cl}^\top PA_{cl} + Q + K^\top RK = P$. Then, the stochastic MPC objective function, which is to be minimized at time k with given x_k, is defined as

$$J_N(x_k, \mathbf{u}_{N|k}) = \mathbb{E}_k\left\{\sum_{l=0}^{N-1}\left(x_{l|k}^\top Qx_{l|k} + u_{l|k}^\top Ru_{l|k}\right) + x_{N|k}^\top Px_{N|k}\right\}.$$

[1] To reduce the conservatism induced by using parameterized feedback policies, a different parameterization that is affine in the disturbance could be used without changing the following results, e.g., fully parameterized disturbance feedback as proposed in, e.g., (van Hessem, 2004; Goulart et al., 2006).

With the above introduction of the nominal system state $z_{l|k}$ and decision variable $v_{l|k}$, the expected value can be computed explicitly, leading to the equivalent deterministic, convex, quadratic cost function

$$J_N(x_k, \mathbf{v}_{N|k}) = \sum_{l=0}^{N-1} \left(z_{l|k}^\top Q z_{l|k} + v_{l|k}^\top R v_{l|k} \right) + z_{N|k}^\top P z_{N|k} + c \tag{3.4}$$

where $c = \mathbb{E}_k \left\{ \sum_{l=0}^{N-1} e_{l|k}^\top (Q + K^\top RK) e_{l|k} + e_{N|k}^\top P e_{N|k} \right\}$ is a constant term which can be neglected in the optimization.

Summarizing, the prototype finite horizon optimal control problem to be solved online is given in the following definition, where the constraint sets $\mathbb{Z}_l$, $\mathbb{V}_l$, and $\mathbb{Z}_f$ are to be derived from the original constraints (3.2) and some suitable terminal constraint as described in the next section.

Definition 3.1 (Simplified finite horizon optimal control problem). *Given the nominal dynamics* (3.3a), *the cost* (3.4), *and the constraint sets* $\mathbb{Z}_l$, $\mathbb{V}_l$, *and* $\mathbb{Z}_f$, *the simplified SMPC finite horizon optimal control problem and optimal value function are given by*

$$V_N(x_k) = \min_{\mathbf{v}_{N|k}} J_N(x_k, \mathbf{v}_{N|k}) \tag{3.5a}$$

$$\text{s.t. } z_{l+1|k} = Az_{l|k} + Bv_{l|k}, \quad z_{0|k} = x_k \tag{3.5b}$$

$$z_{l|k} \in \mathbb{Z}_l, \quad \forall l \in \mathbb{N}_1^N \tag{3.5c}$$

$$v_{l|k} \in \mathbb{V}_l, \quad \forall l \in \mathbb{N}_0^{N-1} \tag{3.5d}$$

$$z_{N|k} \in \mathbb{Z}_f. \tag{3.5e}$$

Similar to the previous chapter, given x_k and a minimizer $\mathbf{v}_{N|k}^*$ of the optimal control problem (3.5), the SMPC control law applied to system (3.1) is defined by $\kappa(x_k) = v_{0|k}^*$. The set of feasible decision variables of the SMPC optimization (3.5) for a given state x_k is denoted by

$$\mathbb{D}_N(x_k) = \{\mathbf{v}_{N|k} \in \mathbb{R}^{mN} \mid \exists\, \mathbf{z}_{N|k} \text{ satisfying } (3.5b) - (3.5e)\}$$

and the set of feasible initial conditions by $\mathbb{X}_N = \{x \in \mathbb{R}^n \mid \mathbb{D}_N(x) \neq \emptyset\}$.

The main goal is to suitably design the nominal constraint sets $\mathbb{Z}_l$, $\mathbb{Z}_f$, and $\mathbb{V}_l$ of the finite horizon optimal control problem (3.5), such that in closed loop the constraints (3.2) are satisfied, recursive feasibility is ensured, and appropriate convergence results can be established.

3.1.2 Constraint relaxation and recursive feasibility

This section addresses the stochastic MPC synthesis part. First, the constraints are relaxed in order to achieve less conservative results and allow for different approximations. Based thereon, the deterministic constraint sets $\mathbb{Z}_l$ and $\mathbb{V}_l$ for the nominal system are derived such that their satisfaction by the nominal system is equivalent to satisfaction of the relaxed chance constraints for the original system (3.1). Finally, these constraint sets are further modified to provide stochastic stability guarantees and recursive feasibility under all admissible disturbance sequences.

We discuss the difference between existence of an a priori unknown feasible solution and feasibility of an a priori known candidate solution, which is unique to stochastic MPC and plays a crucial role in the constraint relaxation and in proving stability. An alternative constraint relaxation is presented, where the probability of a given candidate solution being infeasible is a design parameter. The section concludes with the resulting SMPC algorithm.

Constraint relaxation

In the chance constraints (2.7) the probability distribution l steps ahead given *any* realization in the first $l-1$ steps is considered. While, for the sake of forward invariance of $\mathbb{X}_N$, this was necessary in the case of optimization over feedback functions, it becomes a conservative sufficient condition in the case of optimization over input values. In the following, we take advantage of the probabilistic nature of the disturbance and require that the suboptimal combination of SMPC feed-forward input sequence and static error feedback remains feasible for most, but not necessarily for all possible disturbance sequences. This is in line with the fact that at each sampling time the optimal input is recomputed resulting in a local approximation based on the actual disturbance realization and ensuring that the original constraints (3.2) are satisfied in closed loop.

Let $\varepsilon_u \in (0,1)^N$ be a user defined probabilistic level. In the following, the state and input constraints (3.2) are relaxed *in the optimization* to

$$\mathbb{P}\{[H]_j x_{l|k} \leq [h]_j \mid x_k\} \geq 1 - [\varepsilon]_j \qquad \forall j \in \mathbb{N}_1^p,\ l \in \mathbb{N}_1^N, \tag{3.6a}$$

$$\mathbb{P}\{[G]_j u_{l|k} \leq [g]_j \mid x_k\} \geq 1 - [\varepsilon_u]_j \qquad \forall j \in \mathbb{N}_1^q,\ l \in \mathbb{N}_0^{N-1}. \tag{3.6b}$$

In the SMPC literature, this constraint formulation was first proposed in (Schwarm and Nikolaou, 1999) and is commonly used by now. The main drawback is that it leads to a lack of recursive feasibility, which is discussed in the following. As pointed out in (Korda et al., 2011), one advantage is that compared to (3.2), this leads to lower average costs if the optimal solution is "near" a chance constraint and, as will be shown later, it results in a much larger feasible set $\mathbb{X}_N$. Furthermore, this has computational advantages since various approximation techniques can be used, ranging from the scenario approach (Calafiore and Campi, 2006), deterministic sampling (Lucia et al., 2013) to general function approximation techniques (Mesbah, 2016). The advantage is due to the fact that, unlike in (2.7), the chance constraint on the predicted states $x_{l+1|k}$ need not be satisfied for every realization x_{l+k} of $x_{l|k}$ but only given the current, measured state x_k. The former is essentially equivalent to constraint satisfaction with l worst-case and one stochastic prediction for each prediction horizon $l+1$ and hence the result of using (2.7) is close to what is to be expected for robust MPC.

Derivation of nominal constraints

State and input constraints Given the evolution of the error sequence (3.3b), the relaxed probabilistic constraints (3.6) can equivalently be written in terms of linear constraints on the nominal state $z_{l|k}$ and input $v_{l|k}$ as shown in the following proposition.

Proposition 3.2 (Nominal constraints). *The system* (3.1) *satisfies the chance constraints* (3.6) *if and only if the nominal system* (3.3a) *satisfies the constraints* $z_{l|k} \in \mathbb{Z}_l$ *and* $v_{l|k} \in \mathbb{V}_l$ *with*

$$\mathbb{Z}_l = \{z \in \mathbb{R}^n \mid Hz \leq \eta_l\} \tag{3.7a}$$

$$\mathbb{V}_l = \{v \in \mathbb{R}^m \mid Gv \leq \mu_l\} \tag{3.7b}$$

where η_l *for* $l \in \mathbb{N}_1^N$ *is given by*

$$\begin{aligned} [\eta_l]_j = \max_{\eta}\ & \eta \\ \text{s.t.}\ & \mathbb{P}_k\left\{\eta \leq [h]_j - [H]_j e_{l|k}\right\} \geq 1 - [\varepsilon]_j \end{aligned} \tag{3.8}$$

and μ_l *for* $l \in \mathbb{N}_0^{N-1}$ *is given by*

$$\begin{aligned} [\mu_l]_j = \max_{\mu}\ & \mu \\ \text{s.t.}\ & \mathbb{P}_k\left\{\mu \leq [g]_j - [G]_j K e_{l|k}\right\} \geq 1 - [\varepsilon_u]_j. \end{aligned} \tag{3.9}$$

Proof. We prove the claim for the state constraints. The proof for the input constraints is, *mutatis mutandis*, the same. The constraint (3.6a) can be rewritten in terms of $z_{l|k}$ and $e_{l|k}$ as

$$\mathbb{P}_k\left\{[H]_j z_{l|k} \leq [h]_j - [H]_j e_{l|k}\right\} \geq 1 - [\varepsilon]_j \tag{3.10}$$

with $e_{l|k}$ defined in (3.3b). Constraint (3.10) is equivalent to $\exists \tilde{\eta} \in \mathbb{R}$ s.t. $[H]_j z_{l|k} \leq \tilde{\eta}$ and $\mathbb{P}_k \left\{ \tilde{\eta} \leq [h]_j - [H]_j e_{l|k} \right\} \geq 1 - [\varepsilon]_j$, which is equivalent to $[H]_j z_{l|k} \leq [\eta_l]_j$, with $[\eta_l]_j = \max_{\tilde{\eta}} \tilde{\eta}$ s.t. $\mathbb{P}_k \left\{ \tilde{\eta} \leq [h]_j - [H]_j e_{l|k} \right\} \geq 1 - [\varepsilon]_j$. The maximum value exists as (3.8) can equivalently be written as

$$
\begin{aligned}
-[\eta_l]_j = \min_{\eta}\ & \eta \\
\text{s.t. } & \mathbb{P}_k([H]_j e_{l|k} - [h]_j \leq \eta) \geq 1 - [\varepsilon]_j.
\end{aligned}
$$

By Assumption 3.1 on the disturbance, the cumulative density function F_{He-h} for the random variable $[H]_j e_{l|k} - [h]_j$ exists and is right-continuous. Using F_{He-h}, the constraint can be written as $F_{He-h}(\eta) \geq 1 - [\varepsilon]_j$ which concludes the proof. ∎

The random variable $e_{l|k}$ does neither depend on the realization of the state x_k at time k nor on the optimization variables $\mathbf{v}_{N|k}$, $\mathbf{z}_{N|k}$. Hence these tightened, nominal constraints can be computed by solving offline $(p+q)N$ independent, one dimensional, linear chance constrained optimization programs, leading to an online optimization that is no more difficult to implement than conventional linear MPC. Computational issues of solving the stochastic optimization problems (3.8), (3.9) will be addressed in Section 3.1.4.

Terminal constraint To derive a terminal constraint satisfying the assumptions stated in Chapter 2, we first construct a set which is forward invariant for the closed loop under the local control law $\kappa_f(x) = Kx$ and then employ a suitable tightening to determine the terminal constraint $\mathbb{Z}_f$ for the nominal system.

Proposition 3.3 (Terminal constraint). *For system* (3.1) *with control law* $u_k = Kx_k$ *let* $\mathbb{X}_f = \{x \in \mathbb{R}^n \mid H_f x \leq h_f\}$ *be a robust positively invariant polytope such that* $GKx \leq g$ *and* $HA_{cl}x \leq \eta_1$ *for all* $x \in \mathbb{X}_f$ *and* η_1 *according to* (3.8).

If $\mathbb{X}_f$ *is non-empty, Assumption 2.6 is satisfied with local control law* $\kappa_f(x) = Kx$ *and* $c_1 = \mathbb{E}_w\{\|B_w W_k\|_P^2\}$.

Remark 3.4. In Proposition 3.3 robust positive invariance refers to the disturbance bound $w_k \in \mathbb{W}$, ignoring the stochastic description. For an in depth theoretical discussion, practical computation and polytopic approximations of robust invariant sets see (Blanchini, 1999) for an overview or (Kolmanovsky and Gilbert, 1998) for details.

Proof. With $\kappa_f(x) = Kx$ the hard input constraint is satisfied by definition since $GKx \leq g\ \forall x \in \mathbb{X}_f$. Similarly, the chance constraint on the state is satisfied which follows from Proposition 3.2 since under the local control law $z_{1|k} = A_{cl}x_k$.

Furthermore, by definition, the set $\mathbb{X}_f$ is forward invariant for all disturbances and

$$\begin{aligned}&\mathbb{E}_w\{\|A_{cl}x + B_w W\|_P^2\} + \|x\|_Q^2 + \|Kx\|_R^2\\ =&\|x\|_{A_{cl}^\top PA_{cl}+Q+K^\top RK} + \mathbb{E}_w\{\|B_w W\|_P^2\}\\ =&\|x\|_P^2 + \mathbb{E}_w\{\|B_w W\|_P^2\}\end{aligned}$$

which proves part iii) of Assumption 2.6. ■

In order to define the terminal set $\mathbb{Z}_f$ for the nominal system, a constraint tightening approach similar to (3.8) is employed and the hard terminal constraint is relaxed to a probabilistic constraint. Given a probabilistic level $\varepsilon_f \in [0, 1)$ let

$$\begin{aligned}[\eta_f]_j = &\max_{\eta} \eta\\ &\text{s.t. } \mathbb{P}_k\left\{\eta \leq [h_f]_j - [H_f]_j e_{N|k}\right\} \geq 1 - \varepsilon_f.\end{aligned} \tag{3.11}$$

The terminal constraint set is then given by

$$\mathbb{Z}_f = \{z \in \mathbb{R}^n \mid H_f z \leq \eta_f\}. \tag{3.12}$$

Recursive feasibility

Due to the updated information on the realized state at the next sampling time, in particular since

$$\mathbb{P}\{x_{l|k} \in \mathbb{X} \mid x_k\} \geq 1 - \varepsilon \not\Rightarrow \mathbb{P}\{x_{l|k} \in \mathbb{X} \mid x_k, w_k\} \geq 1 - \varepsilon,$$

the constraint relaxation might lead to infeasibility of the trajectory and infeasibility of the MPC optimization program in general. This situation is depicted in Figure 3.1: While the support of the pdf of the predicted states at time $k+1$ is a subset of the support of the pdf of the predicted states at time k, the pdf itself and thus the probability of constraint satisfaction changes and leads to a violation of the chance constraint.

In the following, a hybrid strategy is proposed: In addition to the relaxed stochastic constraints (3.6) on the predicted states and inputs, we explicitly impose a viability constraint on the applied input $u_{0|k}$. Jointly with the terminal cost and constraint, the former allows to derive stability guarantees while the latter ensures recursive feasibility. At the cost of further offline reachability and controllability set computations, the proposed approach has the advantage of being less conservative compared to the original constraints, but yet guarantees to stabilize the system at the minimal positively invariant region.

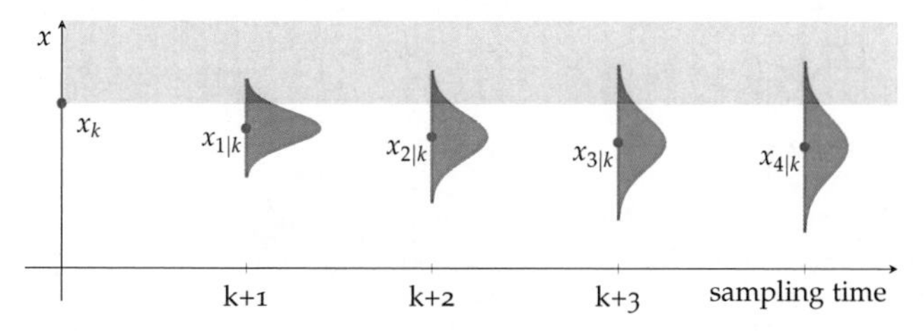

(a) Probability density functions of the states predicted at time k with 0.1 probability of constraint violation (red) given the state x_k.

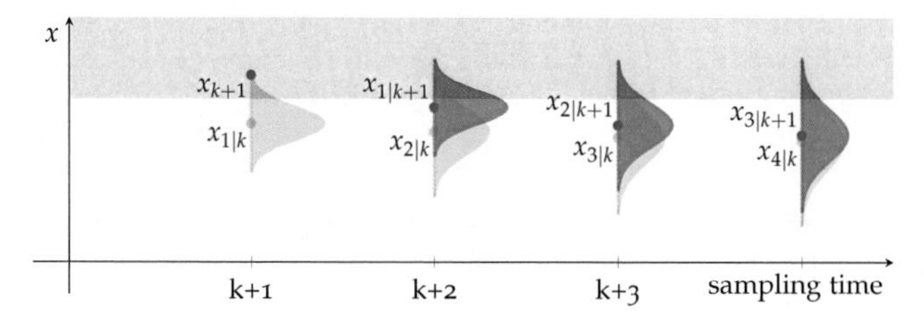

(b) Probability density functions of the predicted states with the same inputs but predicted at time $k+1$, i.e., given the realization of the state x_{k+1}. Due to the additional information, the probability of constraint violation is larger than 0.1.

Figure 3.1. Given a new measurement, the support of the pdf of the predicted states is a subset of the support of the previously predicted pdf, but probabilities change and therefore the previously satisfied chance constraint is violated and makes the trajectory infeasible. In particular, as $\mathbb{P}\{x_{l|k} \in \mathbb{X} \mid x_k\} \geq 1-\varepsilon \not\Rightarrow \mathbb{P}\{x_{l|k} \in \mathbb{X} \mid x_k, w_k\} \geq 1-\varepsilon$ stochastic MPC with chance constraints (3.6) is not recursively feasible.

Recovering recursive feasibility The N-step set and feasible first inputs for the nominal system (3.3a) under the tightened constraint sets $\mathbb{Z}_l$, $\mathbb{V}_l$, and $\mathbb{Z}_f$ is given by

$$\tilde{\mathbb{X}}_N^u = \left\{ (z_{0|k}, v_{0|k}) \in \mathbb{R}^n \times \mathbb{R}^m \;\middle|\; \begin{array}{l} \exists v_{1|k}, \dots, v_{N-1|k} \in \mathbb{R}^m, \\ \mathbf{z}_{N|k} \in \mathbb{R}^{Nn} \text{ satisfying} \\ z_{l+1|k} = Az_{l|k} + Bv_{l|k} \\ z_{l|k} \in \mathbb{Z}_l, \ \forall l \in \mathbb{N}_1^N \\ v_{l|k} \in \mathbb{V}_l, \ \forall l \in \mathbb{N}_0^{N-1} \\ z_{N|k} \in \mathbb{Z}_f \end{array} \right\}.$$

The set $\tilde{\mathbb{X}}_N^u$ defines the feasible states and first inputs of the finite horizon optimal control problem with relaxed constraints. As discussed in the previous paragraph, the projection onto the first n coordinates $\tilde{\mathbb{X}}_N = \text{Proj}_x(\tilde{\mathbb{X}}_N^u)$ is not necessarily robust positively invariant with respect to the disturbance set $\mathbb{W}$. Let $\mathbb{X}_N^\infty$ be a robust control invariant polytope for system (3.1) with state and input satisfying the constraint $(x, u) \in \tilde{\mathbb{X}}_N^u$. A non-conservative solution to recover recursive feasibility of the relaxed stochastic MPC optimization is to explicitly restrict the first predicted state to robustly satisfy the constraint $x_{1|k} \in \mathbb{X}_N^\infty$.

Note that it is important to keep the constraint on the input u in the computation of the robust control invariant set in order to guarantee existence of an input that renders the set robust forward invariant *and* satisfies the MPC constraints.

Remark 3.5. The computation of the sets $\tilde{\mathbb{X}}_N^u$ and $\mathbb{X}_N^\infty$ is a long-standing problem in (linear) controller design, which has gained renewed attention in the context of robust MPC. Efficient algorithms to exactly calculate or to approximate those sets exist, e.g., (Kolmanovsky and Gilbert, 1998; Blanchini and Miani, 2015).

The set $\tilde{\mathbb{X}}_N^u$ can be computed explicitly via projection or iteratively by a backward recursion. To determine the set $\mathbb{X}_N^\infty$ let $\tilde{\mathbb{X}}_N^0 = \tilde{\mathbb{X}}_N$ and

$$\tilde{\mathbb{X}}_N^{i+1} = \left\{ x \in \tilde{\mathbb{X}}_N^i \;\middle|\; \begin{array}{l} \exists u \in \mathbb{R}^m \text{ satisfying} \\ (x, u) \in \tilde{\mathbb{X}}_N^u, \ Ax + Bu \in \tilde{\mathbb{X}}_N^i \ominus B_w \mathbb{W} \end{array} \right\}.$$

The set $\mathbb{X}_N^\infty$ is defined through $\mathbb{X}_N^\infty = \cap_{i=0}^\infty \tilde{\mathbb{X}}_N^i$ and the basis of a standard algorithm to compute $\mathbb{X}_N^\infty$ is given by recursively computing $\tilde{\mathbb{X}}_N^i$ until $\tilde{\mathbb{X}}_N^i = \tilde{\mathbb{X}}_N^{i+1}$ for some $i \in \mathbb{N}$ which implies $\mathbb{X}_N^\infty = \tilde{\mathbb{X}}_N^i$. The basic idea of the algorithm and convergence analysis of the sequence $\tilde{\mathbb{X}}_N^i$ can be found in, e.g., (Bertsekas, 1972) or (Blanchini and Miani, 2015, Section 5.3). Matlab implementations of those algorithms as part of a toolbox can be found in, e.g., (Herceg et al., 2013) or (Kerrigan, 2000).

Recursive feasibility of the candidate solution To prove asymptotic stability, not only existence of a feasible solution at each time k is of interest, but also the probability of feasibility of an explicitly given *candidate solution* at time $k+1$ will be considered. In particular, if the allowed constraint violation is large, it might be desirable to not only a posteriori evaluate that bound but design the relaxed constraints accordingly. In the following, a refined constraint tightening is introduced, which allows to explicitly bound the probability of the candidate solution being infeasible in the next time step.

Given an admissible input trajectory at time k, a candidate solution for time $k+1$ is given by a "shifted solution" as defined in the following.

Definition 3.6 (Candidate solution). *Given an admissible input trajectory $\mathbf{v}_{N|k}$ for the SMPC optimization* (3.5), *the candidate solution $\tilde{\mathbf{v}}_{N|k+1}$ at time $k+1$ is defined by*

$$\tilde{v}_{l|k+1} = \begin{cases} v_{l+1|k} + KA_{cl}^{l} B_w w_k & l \in \mathbb{N}_0^{N-2} \\ K(z_{N|k} + A_{cl}^{l} B_w w_k) & l = N-1 \end{cases} \tag{3.13}$$

where $B_w w_k = x_{k+1} - z_{1|k}$.

In the following, let $\varepsilon_{\tilde{f}} \in [0,1)$ be the chosen upper bound on the probability that $\tilde{\mathbf{v}}_{N|k+1}$ does not satisfy the constraints, that is for all k we require

$$\mathbb{P}_k\{\tilde{\mathbf{v}}_{N|k+1} \notin \mathbb{D}_N(x_{k+1})\} \leq \varepsilon_{\tilde{f}}.$$

Let $\mathbb{W}_f \subseteq \mathbb{W}$ be a convex $1-\varepsilon_{\tilde{f}}$ confidence region for the disturbance W_k, i.e., $\mathbb{P}\{W_k \in \mathbb{W}_f\} \geq 1-\varepsilon_{\tilde{f}}$. To define the tightened constraints, we replace the right hand side of the inequalities defining the constraint sets $\mathbb{Z}_l$, $\mathbb{V}_l$, and $\mathbb{Z}_f$, given in (3.7) and (3.12), respectively, with suitably chosen vectors $\tilde{\eta}_l$, $\tilde{\mu}_l$, and $\tilde{\eta}_f$. For $l \in \mathbb{N}_1^{N-1}$ and $i \in \mathbb{N}_0^{l-1}$ let

$$[\hat{\eta}_{l,i}]_j = \min_{(w_\kappa)_{\kappa \in \mathbb{N}_1^i} \in \mathbb{W}_f^i} -[H]_j \sum_{\kappa=1}^{i} A_{cl}^{l-\kappa} B_w w_\kappa.$$

The refined right hand side of the nominal state constraint for $l \in \mathbb{N}_1^N$ is then given by

$$[\tilde{\eta}_l]_j = \min_{i \in \mathbb{N}_0^{l-1}} \{[\hat{\eta}_{l,i}]_j + [\eta_{l-i}]_j\}. \tag{3.14}$$

Similarly to the idea of "recursively feasible probabilistic tubes" $\hat{\eta}_{l,i}$ takes into account i worst case disturbances based upon the confidence region $\mathbb{W}_f$. However, as the probabilistic tightening defined in (3.8) might be more restrictive than robust constraint tightening based upon $\mathbb{W}_f$, the additional minimization (3.14) becomes necessary.

Analogous to $\tilde{\eta}_l$, define $\tilde{\mu}_l$ by replacing H, h with GK, g, respectively. For the terminal constraint, the robust and stochastic part of the constraint tightening is given by

$$[\hat{\eta}_{f,i}]_j = \min_{(w_\kappa)_{\kappa \in \mathbb{N}_1^i} \in \mathbb{W}_f^i} -[H_f] \sum_{\kappa=1}^{i} A_{cl}^{N-\kappa} B_w w_\kappa,$$

$$[\eta_{f,i}]_j = \max_{\eta} \eta, \text{ s.t. } \mathbb{P}_k\{\eta \leq [h_f]_j - [H_f]_j e_{i|k}\} \geq 1 - \varepsilon_{\bar{f}}$$

which finally leads to the refined right hand side

$$[\tilde{\eta}_f]_j = \min_{i \in \mathbb{N}_0^N} \{[\hat{\eta}_{f,i}]_j + [\eta_{f,N-i}]_j\}.$$

Proposition 3.7 (Recursive feasibility of the candidate solution). *Let the state, input, and terminal constraints in the MPC optimization* (3.5) *be given by*

$$\begin{aligned} \mathbb{Z}_l &= \{z \in \mathbb{R}^n \mid Hz \leq \tilde{\eta}_l\}, \quad l \in \mathbb{N}_1^N \\ \mathbb{V}_l &= \{v \in \mathbb{R}^m \mid Gv \leq \tilde{\mu}_l\}, \quad l \in \mathbb{N}_0^{N-1} \\ \mathbb{Z}_f &= \{z \in \mathbb{R}^n \mid H_f z \leq \tilde{\eta}_f\} \end{aligned} \tag{3.15}$$

with $\mathbb{Z}_f \subseteq \mathbb{Z}_N$. If $\mathbf{v}_{N|k}$ is an admissible input sequence at time k, then, with probability no smaller than $1 - \varepsilon_{\bar{f}}$, the candidate solution $\tilde{\mathbf{v}}_{N|k+1}$ is admissible at time $k+1$, i.e.,

$$\mathbf{v}_{N|k} \in \mathbb{D}_N(x_k) \implies \mathbb{P}_w\{\tilde{\mathbf{v}}_{N|k+1} \in \mathbb{D}_N(Ax_k + Bv_{0|k} + B_w W_k)\} \geq 1 - \varepsilon_{\bar{f}}.$$

Proof. Let $\tilde{\mathbf{z}}_{N|k+1}$ be the state prediction derived from the candidate solution and initial condition $\tilde{z}_{0|k+1} = x_{k+1} = z_{1|k} + B_w w_k$. With probability $1 - \varepsilon_{\bar{f}}$ it holds $W_k \in \mathbb{W}_f$, hence it suffices to prove the claim for $w_k \in \mathbb{W}_f$.

Assume $w_k \in \mathbb{W}_f$, recursive feasibility of the terminal constraint follows from Proposition 3.3, in particular robust forward invariance of the terminal region. Furthermore $\tilde{z}_{N|k+1} \in \mathbb{Z}_N$ is implied by $\mathbb{Z}_f \subseteq \mathbb{Z}_N$. Constraint satisfaction for the state constraints for $l < N$ follows inductively. Assume $Hz_{l|k} \leq \tilde{\eta}_l$ for $l \in \mathbb{N}_1^N$.

Then

$$
\begin{aligned}
[H\tilde{z}_{l|k+1}]_j &= [Hz_{l+1|k} + HA_{cl}^l B_w w_k]_j \\
&\leq [\tilde{\eta}_{l+1} + HA_{cl}^l B_w w_k]_j \\
&= \min_{i=0,\ldots,l} \left\{ [\eta_{l+1,i}]_j + [H]_j A_{cl}^l B_w w_k \right\} \\
&= \min_{i=0,\ldots,l} \left\{ [\eta_{l+1-i}]_j - \max_{w_\kappa \in \mathbb{W}_f} \left[[H]_j \sum_{\kappa=1}^{i} A_{cl}^{l+1-\kappa} B_w w_\kappa \right] + [H]_j A_{cl}^l B_w w_k \right\} \\
&\leq \min_{i=1,\ldots,l} \left\{ [\eta_{l+1-i}]_j - \max_{w_\kappa \in \mathbb{W}_f} \left[[H]_j \sum_{\kappa=2}^{i} A_{cl}^{l+1-\kappa} B_w w_\kappa \right] \right. \\
&\qquad \left. - \max_{w \in \mathbb{W}_f} \left[[H]_j A_{cl}^l B_w w \right] + [H]_j A_{cl}^l B_w w_k \right\} \\
&\leq \min_{i=0,\ldots,l-1} \left\{ [\eta_{l-i}]_j - \max_{w_\kappa \in \mathbb{W}_f} \left[[H]_j \sum_{\kappa=1}^{i} A_{cl}^{l-\kappa} B_w w_\kappa \right] \right\} = \tilde{\eta}_l.
\end{aligned}
$$

Satisfaction of the input constraints follows by a similar computation, replacing H, η_l, and $\tilde{\eta}_l$ by GK, μ_l, and $\tilde{\mu}_l$, respectively, which concludes the proof. ∎

While the constraints (3.7) are easy to derive, they only allow for an analysis of the maximal probability of infeasibility of the candidate solution. On the other hand, the constraints (3.15) are more cumbersome to compute but the maximal probability $\varepsilon_{\bar{f}}$ is an explicit design parameter. This alternative constraint tightening unifies previous approaches and closes the gap between "recursively feasible probabilistic tubes" (Kouvaritakis et al., 2010) which are recovered with $\varepsilon_{\bar{f}} = 0$ and the "least restrictive" scheme presented in (Korda et al., 2011) where only existence of a solution is considered. The impact of $\varepsilon_{\bar{f}}$ on the convergence and provable average closed-loop cost will be highlighted in the next section and the influence on the size of the feasible region is demonstrated in the numerical example in Section 3.1.4.

3.1.3 Stochastic MPC algorithm and closed-loop properties

In summary, the complexity of the control task, which is due to the stochastic disturbance, has been shifted to the offline design phase. This leads to the clear advantage that the computational complexity of the online part of the proposed stochastic MPC algorithm is similar to that of the basic MPC algorithm presented in Chapter 2. Specifically, only a deterministic, convex linearly constrained quadratic program needs to be solved in each iteration. In the following, we present the complete algorithm and thereafter thoroughly analyze its control theoretic properties. In particular, we highlight the influence of $\varepsilon_{\bar{f}}$, the bound on the probability of the candidate solution being infeasible, and contrast the presented results with the existing literature. The main results of this section are summarized in Theorem 3.9 and Theorem 3.12.

Algorithm 3.8 (Stochastic MPC for linear systems with additive disturbance).
Offline: Determine the nominal constraint sets $\mathbb{Z}_l$, $\mathbb{V}_l$, and $\mathbb{Z}_f$ according to either (3.7) and (3.12), or the refined constraint tightening (3.15). Determine the robust control invariant set $\mathbb{X}_N^\infty$.
Online: For each time step $k = 0, 1, 2, \dots$

1. Measure the current state x_k.
2. Determine the minimizer $\mathbf{v}_{N|k}^*$ of the simplified finite horizon optimal control problem (3.5) with additional first step constraint $x_{1|k} \in \mathbb{X}_N^\infty$

$$
\begin{aligned}
\mathbf{v}_{N|k}^* = \arg\min_{\mathbf{v}_{N|k}} \; & J_N(x_k, \mathbf{v}_{N|k}) && && (3.16a) \\
\text{s.t. } & z_{l+1|k} = A z_{l|k} + B v_{l|k}, && z_{0|k} = x_k && (3.16b) \\
& z_{l|k} \in \mathbb{Z}_l && \forall l \in \mathbb{N}_1^N && (3.16c) \\
& v_{l|k} \in \mathbb{V}_l && \forall l \in \mathbb{N}_0^{N-1} && (3.16d) \\
& z_{N|k} \in \mathbb{Z}_f && && (3.16e) \\
& z_{1|k} \in \mathbb{X}_N^\infty \ominus B_w \mathbb{W}. && && (3.16f)
\end{aligned}
$$

3. Apply the SMPC feedback law $\kappa(x_k) = v_{0|k}^*$.

The set of admissible input sequences of the SMPC optimization (3.16) for a given state x_k is denoted by

$$
\mathbb{D}_N(x_k) = \{\mathbf{v}_{N|k} \in \mathbb{R}^{mN} \mid \exists\, \mathbf{z}_{N|k} \text{ satisfying } (3.16b) - (3.16f)\},
$$

the set of feasible initial conditions by $\mathbb{X}_N = \{x \in \mathbb{R}^n \mid \mathbb{D}_N(x) \neq \emptyset\}$, and the optimal value function by V_N. To simplify the presentation of the results, we make the following standard assumption which is typically satisfied if the constraint sets $\mathbb{X}$ and $\mathbb{U}$ are bounded.

Assumption 3.2 (Bounded constraint set). *The set $\mathbb{X}_N$ is bounded.*

Properties of the proposed stochastic MPC scheme

We first derive a mean-square bound on the state, which highlights the connection to known results using the more restrictive, non-relaxed chance constraints, cf. (Kouvaritakis and Cannon, 2016). Thereafter, we prove asymptotic stability in probability of a robust invariant set, which is a novel result in the stochastic MPC literature and which shows the connection to tube based robust MPC approaches like (Chisci et al., 2001; Mayne et al., 2005). This asymptotic behavior has previously been conjectured and shown in simulations in (Deori et al., 2014).

Mean-square bound As discussed in Chapter 2, due to the persistent excitation through the additive disturbance, it is clear that the system does not converge asymptotically to the origin, but ultimately "oscillates" with bounded variance around it. Analogous to Proposition 2.4 and Corollary 2.5, this is made mathematically precise in the following theorem which summarizes the main closed-loop properties of the proposed simplified stochastic MPC algorithm.

Theorem 3.9 (Simplified SMPC closed-loop properties). *Consider the simplified stochastic MPC Algorithm 3.8 in closed loop with the linear system 3.1 and suppose Assumptions 3.1 and 3.2 are satisfied. The SMPC optimization 3.16 is robust recursively feasible, that is $x_k \in \mathbb{X}_N$ implies $Ax_k + B\kappa(x_k) + B_w w_k \in \mathbb{X}_N$ for every realization $w_k \in \mathbb{W}$. For $x_0 \in \mathbb{X}_N$, the resulting state and input trajectories satisfy the hard input and probabilistic state constraints* (3.2) *for all $k \in \mathbb{N}_{>0}$ and*

$$\lim_{t\to\infty} \frac{1}{t} \sum_{k=0}^{t} \mathbb{E}_0 \left\{ \|x_k\|_Q^2 \right\} \le c_1 = (1-\varepsilon_{\bar{f}})\mathbb{E}_w \left\{ \|B_w W_0\|_P^2 \right\} + \varepsilon_{\bar{f}} c_f$$

with $\varepsilon_{\bar{f}}$ being the maximum probability that the candidate solution is not feasible, $c_f = L \max_{w\in\mathbb{W}} \|B_w w\|$, and L the Lipschitz constant of the optimal value function V_N on the set $\mathbb{X}_N$.

Proof. Recursive feasibility follows from the first step constraint as $z_{1|k} \in \mathbb{X}_N^\infty \ominus B_w\mathbb{W}$ implies $x_{k+1} \in \mathbb{X}_N^\infty$ which by construction is a subset of the feasible set.

With $v_{0|k}^* \in \mathbb{V}_0$ and $z_{1|k}^* \in \mathbb{Z}_1$ for all $k \in \mathbb{N}$, chance constraint satisfaction follows from Proposition 3.2 and hard input constraint satisfaction from $e_{0|k} = 0$ and hence $\mu_0 = g$ and thereby $\mathbb{V}_0 = \mathbb{U}$.

To prove the second part, analogously to the proof of Proposition 2.4, we use the optimal value function of (3.16) as a stochastic Lyapunov function. Yet we need to explicitly take into account possible infeasibility of the candidate solution. V_N is known to be continuous, convex, and piecewise quadratic (Bemporad et al., 2002).

Hence, a Lipschitz constant L on $\mathbb{X}_N$ exists and the difference $V(x_{k+1}) - V(z_{1|k})$ can be upper bounded by $L \max_{w \in \mathbb{W}} \|B_w w\|$.

Let $\mathbb{E}\left\{V_N(x_{k+1}) \mid x_k, \tilde{\mathbf{v}}_{N|k+1} \text{ feasible}\right\}$ be the expected optimal value at time $k+1$, conditioning on the state at time k and feasibility of the candidate solution $\tilde{\mathbf{v}}_{N|k+1}$ as defined in (3.13). Then

$$\begin{aligned}
&\mathbb{E}\left\{V_N(x_{k+1}) \mid x_k, \tilde{\mathbf{v}}_{N|k+1} \text{ feasible}\right\} - V_N(x_k) \\
&\leq \sum_{l=1}^{N-1} \left(\|z^*_{l|k}\|_Q^2 + \|v^*_{l|k}\|_R^2\right) + \|z^*_{N|k}\|^2_{(Q+K^\top RK)} + \|z^*_{N+1|k}\|_P^2 \\
&\quad + \mathbb{E}_w \left\{\sum_{l=1}^{N} \|A_{cl}^{l-1} B_w W_k\|^2_{(Q+K^\top RK)} + \|A_{cl}^N B_w W_k\|_P^2\right\} \\
&\quad - \left(\sum_{l=0}^{N-1} \left(\|z^*_{l|k}\|_Q^2 + \|v^*_{l|k}\|_R^2\right) + \|z^*_{N|k}\|_P^2\right) \\
&= \|z^*_{N|k}\|^2_{(Q+K^\top RK)} + \|z^*_{N+1|k}\|_P^2 - \|z^*_{0|k}\|_Q^2 - \|v^*_{0|k}\|_R^2 - \|z^*_{N|k}\|_P^2 + \mathbb{E}_w\left\{\|B_w W_k\|_P^2\right\} \\
&\leq -\|z_{0|k}\|_Q^2 + \mathbb{E}_w\left\{\|B_w W_k\|_P^2\right\} = -\|x_k\|_Q^2 + \mathbb{E}_w\left\{\|B_w W_k\|_P^2\right\}
\end{aligned}$$

where $v^*_{l|k}$ for $l = 0, \dots, N-1$ denotes the optimal solution of (3.16) at time k, $z^*_{l|k}$, for $l = 0, \dots, N$ the corresponding state sequence, and $z^*_{N+1|k} = A_{cl} z^*_{N|k}$. Note that the expected value of all w-z cross-terms equals zero because of the zero-mean and independence assumption. Furthermore, since we defined the terminal cost as the solution to the discrete-time Lyapunov equation it holds $A_{cl}^\top P A_{cl} + Q + K^\top RK = P$.

Combining both cases we obtain by the law of total expectation

$$\begin{aligned}
&\mathbb{E}\left\{V_N(x_{k+1}) \mid x_k\right\} - V_N(x_k) \\
&\leq (1-\varepsilon_{\bar{f}})\left(\mathbb{E}\left\{V_N(x_{k+1}) | x_k, \tilde{\mathbf{v}}_{N|k+1} \text{ feasible}\right\} - V_N(x_k)\right) + \\
&\quad \varepsilon_{\bar{f}}\left(-\|x_k\|_Q^2 + L \max_{w \in \mathbb{W}} \|B_w w\|\right) \\
&\leq -\|x_k\|_Q^2 + (1-\varepsilon_{\bar{f}})\mathbb{E}\left\{\|B_w W_k\|_P^2\right\} + \varepsilon_{\bar{f}} c.
\end{aligned}$$

The final statement follows by taking iterated expectations. ■

Remark 3.10. Following the argument in (Ghaemi et al., 2008), Theorem 3.9 remains valid if a terminal region, which is robust forward invariant with respect to a disturbance $w_k \in A_{cl}^N \mathbb{W}$ instead of $w_k \in \mathbb{W}$ is used. Similarly, qualitatively the same results can be derived assuming only forward invariance with probability $\varepsilon_{\bar{f}}$ instead.

Remark 3.11. The same corollaries as in Section 2.2 apply.

Asymptotic stability Let $\mathbb{X}_\infty$ be the minimal robust positively invariant set for the system (3.1) with input $u_k = Kx_k$. In the following, we prove that, under mild assumptions, $\mathbb{X}_\infty$ is asymptotically stable in probability for the closed-loop system with the proposed stochastic MPC algorithm. In other words, the proposed SMPC control law stabilizes the same set as the robust MPC proposed in (Chisci et al., 2001). Similarly, the same result applies for the more restrictive stochastic MPC presented in (Kouvaritakis et al., 2010). However, the different nominal constraints lead to a possibly different transient phase and possibly slower convergence rate.

To streamline the presentation, we make the following assumption on the control gain K, as well as a non-restrictive technical assumption on $\mathbb{X}_\infty$ and $\mathbb{X}_f$.

Assumption 3.3 (Prestabilizing feedback, terminal set).

i) The feedback gain K for the prestabilizing and terminal controller is chosen to be the unconstrained LQ-optimal solution.

ii) Let $\mathbb{B}_\lambda \subset \mathbb{R}^n$ be an open ball of radius λ. There exists $\lambda \in \mathbb{R}_{>0}$ such that $\mathbb{X}_\infty \oplus \mathbb{B}_\lambda \subseteq \mathbb{X}_f$.

Theorem 3.12 (Asymptotic stability). *Consider the simplified stochastic MPC Algorithm 3.8 in closed loop with the linear system (3.1) and suppose Assumption 3.1, 3.2, and 3.3 are satisfied. The set $\mathbb{X}_\infty$ is asymptotically stable in probability with region of attraction $\mathbb{X}_N$.*

Remark 3.13. Assumption 3.3 i) could be replaced by a more technical assumption on the unconstrained optimal solution. In this case Theorem 3.12 remains valid with a different set $\mathbb{X}_\infty$.

In the following, we first prove the theorem under the assumption that the candidate solution remains feasible at each time step. Then, we prove that there exists a set $\mathbb{S}$ where this feasibility assumption is verified and that for every probability $\rho \in (0,1]$ and state x_0 in $\mathbb{X}_N^\infty$ there exists a time $T \in \mathbb{N}$ such that $\mathbb{P}\{x_T \in \mathbb{S}\} \geq 1-\rho$. The proof differs from standard proofs using a stochastic Lyapunov function (Kushner, 1967) because of the nonzero probability that the candidate solution does not remain feasible during a transient phase.

Similar to what is known in robust tube MPC (Chisci et al., 2001, Theorem 8), the following lemma establishes convergence under the assumption that the candidate solution remains feasible at each sampling time.

Lemma 3.14. *Consider the simplified stochastic MPC Algorithm 3.8 for the linear system (3.1). Suppose Assumptions 3.1, 3.3 are satisfied and the candidate solution $\tilde{\mathbf{v}}_{N|k+1}$ is feasible for all $k \in \mathbb{N}_{>0}$. The state $x_k = \zeta_k + \xi_k$ can be separated into a part ζ_k and a part ξ_k such that the origin is asymptotically stable for ζ_k with region of attraction $\mathbb{X}_N$ and $\xi_k \in \mathbb{X}_\infty$ $\forall k \geq 0$.*

Proof. Let

$$\zeta_{k+1} = A_{cl}\zeta_k + B(\kappa(x_k) - K(\zeta_k + \xi_k)), \qquad \zeta_0 = x_0, \qquad (3.17a)$$
$$\xi_{k+1} = A_{cl}\xi_k + B_w w_k, \qquad \xi_0 = 0. \qquad (3.17b)$$

Given feasibility of the candidate solution, it has been shown (Chisci et al., 2001) that $c_k = \kappa(x_k) - K(\zeta_k + \xi_k)$ is bounded and $c_k \to 0$ for $k \to \infty$. Since A_{cl} is Schur stable, the system (3.17a) is input to state stable (ISS) with respect to the input c_k and hence ζ_k converges to the origin. Furthermore, for $x_k \in \mathbb{X}_f$ we have $c_k = 0$ (Kouvaritakis and Cannon, 2016), which together with Assumption 3.3 implies asymptotic stability of the origin for system (3.17a). ■

Corollary 3.15. *Consider the simplified stochastic MPC Algorithm 3.8 in closed loop with the linear system (3.1). Suppose Assumptions 3.1, 3.2, 3.3 are satisfied and the candidate solution $\tilde{\mathbf{v}}_{N|k+1}$ is feasible for all $k \in \mathbb{N}_{>0}$. There exists T_ε such that $\|\zeta_k\| < \varepsilon$ for all $k \geq T_\varepsilon$. In particular, there exists $T_f \in \mathbb{N}$ such that $x_k \in \mathbb{X}_f$ for all $k \geq T_f$ and $x_0 \in \mathbb{X}_N$.*

Proof. From asymptotic stability, it follows that the origin is a uniform attractor for (3.17a) and hence for each $x_0 \in \mathbb{X}_N$ there exists a neighborhood $\mathcal{N}_{x_0}$ of x_0 and a sampling time $T_{x_0} \in \mathbb{N}$ such that $\forall x_0 \in \mathcal{N}_{x_0}$ it holds $\zeta_k \in \mathbb{B}_\lambda$ $\forall k > T_{x_0}$ (Bhatia and Szegő, 1967). The collection $\{\mathcal{N}_{x_0}\}_{x_0 \in \mathbb{X}_N}$ is an open covering of the set $\mathbb{X}_N$, hence by compactness of $\mathbb{X}_N$ we can choose a finite subcollection $\{\mathcal{N}_{x_0}\}_{x_0 \in J}$ of $\{\mathcal{N}_{x_0}\}_{x_0 \in \mathbb{X}_N}$ which also covers $\mathbb{X}_N$. Choosing $T_f = \max_{x_0 \in J} T_{x_0}$ it follows $\zeta_k \in \mathbb{B}_\lambda$ $\forall k > T_f$ and all $x_0 \in \mathbb{X}_N$. Since $\xi_k \in \mathbb{X}_\infty$, by Assumption 3.3 this implies $x_k \in \mathbb{X}_f$ for all $k > T_f$ and $x_0 \in \mathbb{X}_N$. ■

The next lemma shows that the assumption on feasibility of the candidate solution is satisfied for all $k > k'$ if $x_{k'}$ is inside the terminal region. Hence inside $\mathbb{X}_f$ Lemma 3.14 can be applied and we only need to consider $x_0 \notin \mathbb{X}_f$.

Lemma 3.16. *Consider the simplified stochastic MPC Algorithm 3.8 for the linear system (3.1) and suppose Assumptions 3.1, 3.3 are satisfied. The terminal region $\mathbb{X}_f$ is robust forward invariant for the closed-loop system. If $x_{k'} \in \mathbb{X}_f$, then $\tilde{\mathbf{v}}_{N|k+1} \in \mathbb{D}_N(x_{k+1})$ for all $k \geq k'$.*

Proof. The unconstrained optimal solution to (3.16) equals the control inputs generated by the LQR, that is $v^*_{l|k} = Kz^*_{l|k}$ (Kouvaritakis and Cannon, 2016). If $z_{0|k} \in \mathbb{X}_f$ robust forward invariance of the terminal region implies that the unconstrained optimal solution satisfies the constraints as

$$\begin{aligned} & HA_{cl}(z_{l|k} + e_{l|k}) \leq \eta_1 && \forall e_{l|k} \\ \Leftrightarrow\ & Hz_{l+1|k} \leq \eta_1 - HA_{cl}e_{l|k} && \forall e_{l|k} \\ \Rightarrow\ & Hz_{l+1|k} \leq \eta_{l+1} \end{aligned}$$

and similarly for the input and terminal constraints. Hence, in the terminal region, the proposed stochastic MPC control law equals the unconstrained LQR. Since $\mathbb{X}_f$ is robust forward invariant under the unconstrained LQR the statement follows. ∎

Under Assumption 3.3, Lemma 3.14 and 3.16 suffice for stability of the proposed algorithm. To prove attractivity, the following lemma provides a bound on the probability that the candidate solution remains feasible for all sampling times in a given interval. In particular, Lemma 3.17 states, that for any given probability $1 - \rho$ a sufficiently long horizon can be found such that, at some point within this horizon, the candidate solution remains feasible T_f consecutive times. By Corollary 3.15 and Lemma 3.16, this implies that the final state is inside the terminal region.

Lemma 3.17. *Consider the simplified stochastic MPC Algorithm 3.8 for the linear system (3.1) and suppose Assumption 3.1 is satisfied. Let $I_{k'} = [k', k' + T_f - 1]$ denote some interval of length T_f and $\mathcal{A}_{k'}$ the event that the candidate solution $\tilde{\mathbf{v}}_{N|k+1}$ is feasible for all $k \in I_{k'}$. For each $\rho \in (0, 1]$ there exists $T_\rho \in \mathbb{N}$ such that*

$$\mathbb{P}\left\{\cup_{k'=0}^{T_\rho} \mathcal{A}_{k'}\right\} \geq 1 - \rho. \tag{3.18}$$

Proof. Let $1 - \varepsilon_{\tilde{f}}$ denote the probability that the candidate solution remains feasible in the next sampling instant. The left hand side of (3.18) can be crudely overapproximated by the probability of remaining feasible during one of the intervals $I_i = [iT_f, (i+1)T_f - 1]$ for $i \in [0, \lfloor \frac{T_\rho}{T_f} \rfloor]$. For each I_i we have $\mathbb{P}\{\mathcal{A}_{iT_f}\} \geq (1 - \varepsilon_{\tilde{f}})^{T_f} =: 1 - \beta_f$. Hence $\mathbb{P}\left\{\cup_{k'=0}^{T_\rho} \mathcal{A}_{k'}\right\} \geq \mathbb{P}\left\{\cup_{i=0}^{\lfloor \frac{T_\rho}{T_f} \rfloor} \mathcal{A}_{iT_f}\right\} \geq 1 - (\beta_f)^{\lfloor \frac{T_\rho}{T_f} \rfloor + 1}$. Since $\beta_f \in [0, 1)$, the right hand side of the inequality is non decreasing with T_ρ and converges to 1. ∎

Lemma 3.18 (Attractivity). *Consider the simplified stochastic MPC Algorithm 3.8 for the linear system* (3.1) *and suppose Assumptions* 3.1, 3.2, *and* 3.3 *are satisfied. For all* $\varepsilon \in \mathbb{R}_{>0}$

$$x_0 \in \mathbb{X}_N \implies \lim_{k' \to \infty} \mathbb{P}\{\sup_{k>k'} \|x_k\|_{\mathbb{X}_\infty} < \varepsilon\} = 1.$$

By Corollary 3.15 and Lemma 3.16 the closed-loop system converges if the candidate solution remains feasible for N_f consecutive time-steps. By Lemma 3.17, for any given probability ρ, there exists a sufficiently long horizon T_ρ such that this holds with probability ρ. We use the Borel-Cantelli Lemma and Fatou's Lemma (Klenke, 2014) to show that in fact the probability grows *fast enough*.

Proof. Let $T = \max\{T_\varepsilon, T_f\}$ with T_ε, T_f according to Corollary 3.15 and define the event $\mathcal{B}_{k'} = \left\{\sup_{k>k'} \|x_{k+N}\|_{\mathbb{X}_\infty} \geq \varepsilon\right\}$. By Corollary 3.15, Lemma 3.16 it holds $\mathbb{P}\{\mathcal{B}_{k'}\} \leq 1 - \mathbb{P}\left\{\cup_{k=0}^{k'} \mathcal{A}_k\right\}$. Inserting the explicit bound derived in the proof of Lemma 3.17 leads to

$$\sum_{k=0}^{\infty} \mathbb{P}\{\mathcal{B}_k\} \leq \sum_{k=0}^{\infty} \beta_f^{\frac{k}{T_f}} < \infty.$$

By the Borel-Cantelli Lemma we have that $\mathbb{P}\left\{\lim_{k\to\infty} \sup \mathcal{B}_k\right\} = 0$ and hence by Fatou's Lemma $\lim_{k\to\infty} \sup \mathbb{P}\left\{\mathcal{B}_k\right\} = 0$ which concludes the proof. ∎

Proof (Theorem 3.12). Stability follows from 3.14 together with robust recursive feasibility of the candidate solution in the terminal region as shown in Lemma 3.16. Attractivity follows from Lemma 3.18. ∎

A direct corollary of Theorem 3.12 is a tighter bound on the asymptotic average performance.

Corollary 3.19 (Asymptotic bound). *Consider the simplified stochastic MPC Algorithm 3.8 for the linear system* (3.1) *and suppose Assumption* 3.1, 3.2, *and* 3.3 *are satisfied. For* $x_0 \in \mathbb{X}_N$ *the closed-loop trajectories satisfy*

$$\lim_{t\to\infty} \frac{1}{t} \sum_{k=0}^{t} \mathbb{E}_0\left\{\|x_k\|_Q^2\right\} \leq \mathbb{E}_w\left\{\|B_w W_0\|_P^2\right\}.$$

Proof. Let $\rho_f(k)$ be the probability that the candidate solution $\tilde{\mathbf{v}}_{N|k+1}$ is infeasible and $\Delta_V(k') = \mathbb{E}\{V(x_{k'})\} - V(x_0)$. Analogous to the proof of Theorem 3.9 it holds

$$\Delta_V(k') \leq \sum_{k=0}^{k'} \mathbb{E}_0\left\{-\|x_k\|_Q^2 + \|B_w W_k\|_P^2\right\} + \rho_f(k)c_f$$

$$\frac{1}{k'}\sum_{k=0}^{k'} \mathbb{E}_0\left\{\|x_k\|_Q^2\right\} + \frac{\Delta_V(k')}{k'} \leq \mathbb{E}_0\left\{\|B_w W_k\|_P^2\right\} + \frac{c_f}{k'}\sum_{k=0}^{k'} \rho_f(k).$$

Since $\sum_{k=0}^{k'} \rho_f(k) \leq \sum_{k=0}^{k'} \varepsilon_{\bar{f}} \mathbb{P}\{\mathcal{B}_{k-N-1}\} < \infty$ the result follows by taking the limit $k' \to \infty$. ■

Summarizing, we have shown that the proposed stochastic MPC algorithm has less restrictive constraints and thus a larger admissible region and more control authority compared to state of the art stochastic MPC algorithms as presented in, e.g., (Kouvaritakis and Cannon, 2016). At the same time, it shares favorable system theoretic properties, in particular, the same robust invariant set is asymptotically stabilized, a result that has previously been conjectured but not been proven.

3.1.4 Implementation details and numerical example

To complete the computational approach, we briefly review numerical methods to solve the single chance constrained programs which appear in the derivation of the nominal constraints. Thereafter we illustrate our results, especially the non-conservativeness of the approach with respect to the allowed probability of constraint violation and the increased feasible region, in a numerical example.

Solving the chance constrained programs

Numerically solving or approximating the solution of chance constrained programs is a well studied topic. We briefly review deterministic as well as sampling based solutions to efficiently evaluate the stochastic programs (3.8), (3.9), and (3.11).

Deterministic methods As probability is simply a semantic obfuscation for integral, the chance constraints can be written as constraints on multivariate integrals. Similar to the derivations in Chapter 2, if the random variable W_k has a known probability density function $f_W(w)$, we can write (3.8) as

$$\begin{aligned}[\eta_l]_j = \max_{\eta}\ & \eta \\ \text{s.t.}\ & \int_{\mathbb{W}^l} \mathbb{1}_{\left\{\eta \leq [h]_j - [H]_j e_{l|k}\right\}} \prod_{i=0}^{l-1} f_W(w_i) \mathrm{d}w_0 \cdots \mathrm{d}w_{l-1} \geq 1 - [\varepsilon]_j\end{aligned}$$

with $e_{l|k} = \sum_{i=0}^{l-1} A_{cl}^i B_w W_i$ and $\mathbb{1}_{\{\cdot\}}$ being the indicator function. By approximating the integral numerically with quadrature rules suitable for high-dimensional integrals like Quasi-Monte Carlo or Sparse Grid methods, the optimization can be solved with standard nonlinear optimization solvers.

If the individual random variables in the random vector W_k are independent, i.e., $f_W(w) = \prod_{s=1}^{m_w} f_{W_s}([w]_s)$, the constraint can be further simplified to a one-dimensional integral. Let $\tilde{h}^i = [H]_j A_{cl}^i B_w$ and define

$$f_{\tilde{h}_s^i W_s}([w]_s) = \frac{1}{|[\tilde{h}^i]_s|} f_{W_s}\left(\frac{[w]_s}{[\tilde{h}^i]_s}\right),$$
$$f_{\tilde{h}^i} = f_{\tilde{h}_1^i W_1} * f_{\tilde{h}_2^i W_2} * \ldots * f_{\tilde{h}_{m_w}^i W_{m_w}},$$

where $f * g$ denotes the convolution of f and g. With this, the probability density function $f_{H_j e_l}$ of $[H]_j e_{l|k}$ is given by

$$f_{H_j e_l} = f_{\tilde{h}^0} * f_{\tilde{h}^1} * \ldots * f_{\tilde{h}^{l-1}}$$

and

$$\begin{aligned} -[\eta_l]_j = \min_{\eta} \ & \eta \\ \text{s.t.} \ & \int_{-\infty}^{\eta} f_{H_j e_l}(x + [h]_j) \mathrm{d}x \geq 1 - [\varepsilon]_j. \end{aligned}$$

The advantage of this formulation is that only a series of one-dimensional integrals need to be solved which can be easily evaluated numerically. Note that due to the multiple convolutions, it can be beneficial to work with the Fourier Transform of f_{W_s} instead of a direct evaluation.

Further discussions on approximations and tailored numerical optimization algorithms can be found in, e.g., (Prékopa, 2010, Chapter 8).

Stochastic methods In recent years, significant development has been made on the theory of solving chance constrained optimization problems based on a finite number of random samples of the uncertainty, cf. (Tempo et al., 2013; Calafiore et al., 2011). The basic idea is to replace the chance constraints by drawing a sufficiently large number N_s of independent realizations of the random variable and require the constraints to hold for all, but a fixed number r of these realizations. Tight sample bounds for different problem classes have been derived to provide rigorous probabilistic guarantees about the correctness of the approximate solution. The results are generally independent of the underlying distribution and the methods are easy to implement without solving high dimensional integrals or even knowing the probability density function.

Campi and Garatti (2011) derive explicit sample complexities which can be used to approximate the chance constrained programs (3.8), (3.9), and (3.11) by solving a sampling based, deterministic optimization program. With a user chosen confidence level $1-\delta$, the sampling based solution satisfies the chance constraint with $\varepsilon \in [\varepsilon_l, \varepsilon_u]$ if N_s and r are chosen such that

$$\varepsilon_l N_s - 1 + \sqrt{3\varepsilon_l N_s \ln\frac{2}{\delta}} \leq r \leq \varepsilon_u N_s - \sqrt{2\varepsilon_u N_s \ln\frac{1}{\delta}}. \tag{3.19}$$

In general, the sample approximation leads to mixed integer problems or heuristics need to be used to discard samples in the optimization. As summarized in the following proposition, here a sort algorithm suffices to find an exact solution.

Proposition 3.20 (Sampling based approximation). *Let $N_s, r \in \mathbb{N}_{>0}$ be chosen according to (3.19) and $\varepsilon_u N_s > k$. Let $e_{l|k}^{(i)} = \sum_{j=1}^{l} A_{cl}^{j-1} B_w w_j^{(i)}$ be independently chosen samples from W^l and let $q_{1-\frac{r}{N_s}}$ be the $(1-\frac{r}{N_s})$-quantile of the set $\left\{[H]_j e_{l|k}^{(i)}\right\}_{i=1,\dots,N_s}$. With confidence $1-\delta$, $[\eta_l]_j = [h]_j - q_{1-\frac{r}{N_s}}$ solves (3.8) for some $\varepsilon \in [\varepsilon_l, \varepsilon_u]$.*

Remark 3.21. Proposition 3.20 holds similarly for (3.9) and (3.11) by replacing H, h with GK, g and H_f, h_f, respectively.

Remark 3.22. If the nominal constraints are derived using the described sampling approach, the chance constraints are only satisfied with confidence $(1-\delta)^p$. Furthermore, due to the additional uncertainty, the closed loop becomes a partially observable Markov Chain with additional state variables and the constraint satisfaction results do not hold if probabilities conditioning on the sequence $x_0, \dots, x_k$ instead of only x_k are considered. However, the MPC optimization is recursively feasible independent of the method used to derive the nominal constraints.

Numerical example

In the following, the performance of the proposed simplified stochastic MPC algorithm will be assessed by means of a numerical example. In particular, the issues of constraint satisfaction and feasible region will be addressed specifically.

Consider the linear system (3.1) with

$$A = \begin{bmatrix} 1 & 0.0075 \\ -0.143 & 0.996 \end{bmatrix}, \quad B = \begin{bmatrix} 4.798 \\ 0.115 \end{bmatrix}, \quad B_w = I_2.$$

The disturbance distribution is assumed to be a truncated Gaussian distribution with the covariance matrix $\Sigma = 0.04^2 I_2$ truncated at $\|w\|^2 \leq 0.02$. The example is a linearized model of a DC-DC converter, which has previously been used in different publications on stochastic MPC, e.g., (Cannon et al., 2011). The system is subject to a single chance constraint on the inductor current with violation probability $\varepsilon = 0.8$

$$\mathbb{P}_k \left\{ [1\ 0] x_{1|k} \leq 2 \right\} \geq 0.8. \tag{3.20}$$

As in (Cannon et al., 2011), the SMPC cost weights were chosen to be $Q = \text{diag}(1, 10)$, $R = 1$ and the prediction horizon to be $N = 8$. The prestabilizing gain K was set to the unconstrained LQR.

For the results in the following paragraph, the nominal constraints are based upon (3.7) and (3.12), i.e., without explicitly bounding $\varepsilon_{\bar{f}}$, the probability of the candidate solution not being feasible. The offline optimizations were performed with the described sampling approach with $\varepsilon_l = 0.95\varepsilon$, $\varepsilon_u = 1.05\varepsilon$ and confidence $\delta = 10^{-4}$. For the robust set calculations a polytopic outer approximation with 8 hyperplanes has been used and the algorithms presented in (Kerrigan, 2000).

The simulations were performed with Matlab using the standard interior point algorithm of `quadprog`. The reported computation time to solve the online optimization in non-condensed form without any further preconditioning was approximately 4ms on an Intel Core i7 with 3.4GHz.

Constraint satisfaction For the initial state $x_0 = [2.5\ 2.8]^\top$, under the stochastic MPC algorithm presented in (Cannon et al., 2011) constraint violation with a fraction of 14.4% in the first 6 steps has been reported. Although the proposed algorithm did not make full use of the allowed violation probability, a lower closed-loop cost compared to robust MPC has been reported.

In contrast, the stochastic MPC algorithm derived in this chapter makes full use of the allowed violation probability leading to an even lower closed-loop cost. In a Monte Carlo simulation with 10^4 realizations of the disturbance sequence an average constraint violation of 20% during the first 6 iterations has been observed. Simulation results of state sequences with 200 random disturbance realizations are shown in Figure 3.2. The left plot shows the complete trajectories for a simulation time of 15 time steps. The right plot shows the constraint violation during the first 6 sampling times in more detail. Constraint (3.20) is satisfied non-conservatively thereby leaving more control authority for optimizing the performance.

Finally, for comparison, we remark that the unconstrained LQ optimal solution violates the constraint 100% in the first 3 steps, nominal MPC violates the constraints with a probability of 0.5, and robust MPC achieves 0% constraint violation.

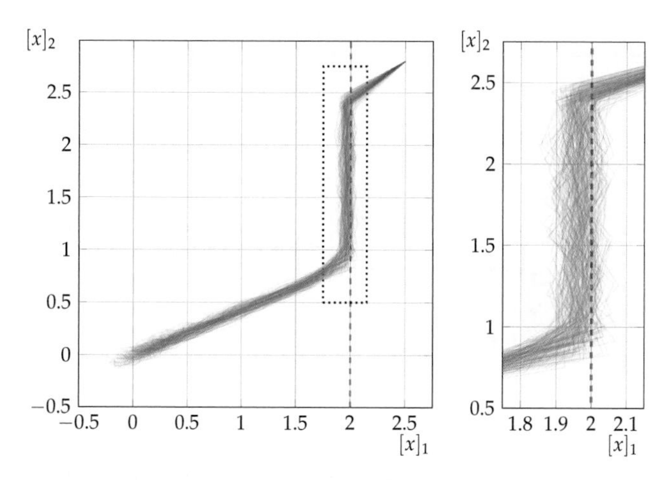

Figure 3.2. Closed-loop response for initial condition $x_0 = [2.5\ 2.8]^\top$ and 200 different disturbance realizations. The detail on the right shows the trajectories near the chance constraint $\mathbb{P}_k\{[x_{1|k}]_1 \leq 2\} \geq 0.8$. The allowed violation probability of $\varepsilon = 0.2$ is fully exploited during the first 6 iterations as confirmed in a Monte Carlo simulation with 10^4 realizations.

Feasible region As highlighted earlier, the main advantage of the proposed stochastic MPC using relaxed constraints is the reduced conservatism leading to a significant increase of the feasible region. To illustrate this fact, in the following, additional chance constraints and input constraints

$$\begin{aligned} &\mathbb{P}_k\{\ [x_{1|k}]_1 \leq 2\ \} \geq 0.8, \quad \mathbb{P}_k\{\ -[x_{1|k}]_1 \leq 2\ \} \geq 0.8, \\ &\mathbb{P}_k\{\ [x_{1|k}]_2 \leq 3\ \} \geq 0.8, \quad \mathbb{P}_k\{\ -[x_{1|k}]_2 \leq 3\ \} \geq 0.8, \\ &\qquad |u| \leq 0.2 \end{aligned}$$

are considered. The nominal constraints for the predicted input and terminal state were computed using relaxed chance constraints with $\varepsilon_u = \varepsilon_{\bar{f}} = 0.05$ as described in Section 3.1.2.

The area of the admissible region $\mathbb{X}_N$ of the proposed stochastic MPC with horizon length $N = 8$ is 1.7 times larger than the area of the admissible region of the stochastic MPC algorithm using non-relaxed constraints. It is 3.4 times larger than the classical robust MPC algorithm presented in (Mayne et al., 2005), which, due to the robust constraints, is of course significantly smaller. Figure 3.3 shows the different admissible regions of the proposed stochastic MPC using relaxed chance constraints (3.7) in combination with the first step constraint, the stochastic MPC with non-relaxed "recursively feasible probabilistic tubes" and the robust MPC algorithm.

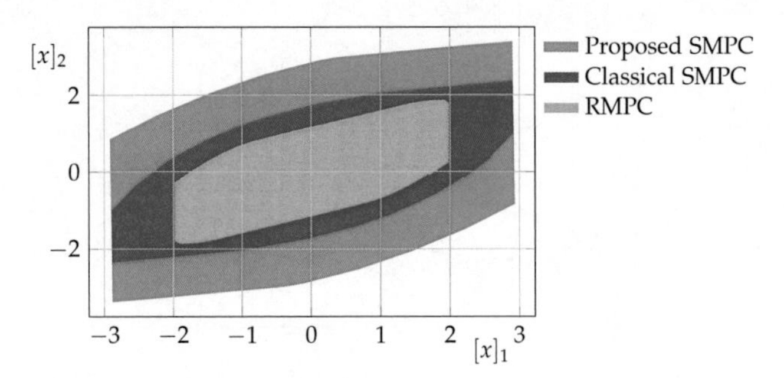

Figure 3.3. Comparison of the admissible region $\mathbb{X}_N$ for the proposed stochastic MPC with relaxed constraints and guaranteed recursive feasibility, stochastic MPC with recursively feasible probabilistic tubes, and robust MPC.

To further highlight the advantage of using relaxed constraints and to show the connection to recursively feasible tubes, in Figure 3.4 the area of the admissible region is plotted against $\varepsilon_{\bar{f}}$ for the case that the proposed constraint tightening with a guaranteed probability of feasibility of the candidate solution according to Section 3.1.2 is employed. It can be observed that even for moderate values of $\varepsilon_{\bar{f}}$ a significant increase in $\mathbb{X}_N$ can be obtained and for $\varepsilon_{\bar{f}} \to 1$ the area of the admissible region is equivalent to using recursively feasible tubes.

3.1.5 Discussion and summary

To conclude this section, we discuss the presented stochastic MPC algorithm with a particular focus on deriving nominal constraints and the assumption of single chance constraints.

Offline simplification of joint chance constraints

It is a known result in robust tube MPC for linear systems, that linear constraints which should be satisfied robustly can equivalently be rewritten as constraints on the nominal state and input. In particular, the nominal constraint set $\mathbb{Z}_l$ involves

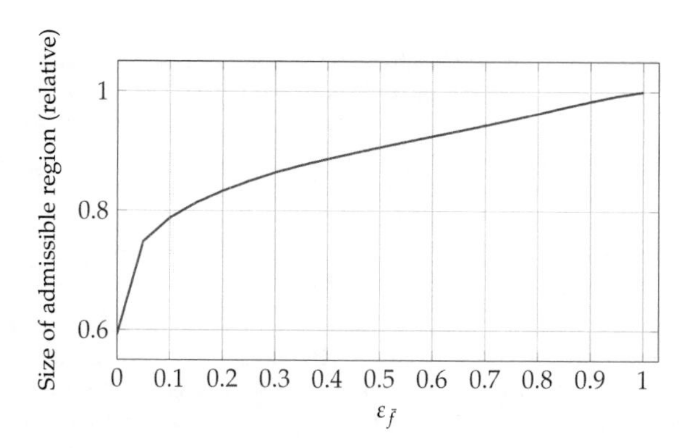

Figure 3.4. Relative size of the admissible region $\mathbb{X}_N$ plotted over $\varepsilon_{\bar{f}}$, the upper bound on the probability of the candidate solution being infeasible. The area is normalized with respect to the case $\varepsilon_{\bar{f}}$.

at most as many linear inequalities as the original constraint set (Rawlings et al., 2017).

We have shown that this holds true for the single chance constraints (3.2a), yet this fact does not extend to joint chance constraints

$$\mathbb{P}_k\{Hx_{1|k} \leq h\} \geq 1 - \varepsilon \tag{3.21}$$

with a number of linear inequalities $p > 1$. An equivalent nominal constraint set $\mathbb{Z}_l$ can in general not only not be expressed by a finite number of linear constraints but it can even be a non-convex set. Even in examples where $\mathbb{Z}_l$ is given by a polytope, e.g., polytopic disturbance set $\mathbb{W}$ with uniform distribution, the nominal constraint is of higher complexity than the original one. An example is given in Figure 3.5 where, depending on the distribution of the uncertainty, the nominal constraint set is not polytopic or of higher complexity. In order to not increase the complexity, single chance constraints (3.2a) can be used to approximate joint chance constraints (3.21), at the cost of being either more conservative or increase the probability of constraint violation at certain points in the state space

A different approach often proposed, e.g., (Cannon et al., 2011; Zhang et al., 2013; Hewing and Zeilinger, 2018) is to determine a confidence region for the disturbance W_k or the prediction error $e_{l|k}$ and then requiring the constraints to hold for all realizations within these sets. Similar to an approximation with single chance constraints, this approach does not approximate the true joint chance con-

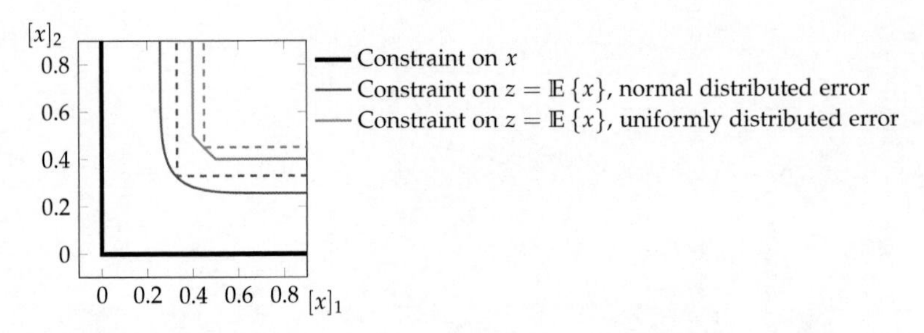

Figure 3.5. Constraint set $x \geq 0$ (solid black) and nominal constraint sets for $z = \mathbb{E}\{x\}$ which are equivalent to a joint chance constraint with violation probability $\varepsilon = 0.1$ for $e = x - z$ being a Gaussian distribution with covariance matrix $\Sigma = 0.2^2 I_2$ (blue) or uniform distribution $e \sim \mathcal{U}([-0.5, 0.5]^2)$ (orange). Dashed lines show the nominal constraints for single chance constraints with $[\varepsilon]_j = \varepsilon/2$.

straint but is equivalent to tightening the original single constraints. Furthermore, choosing a parameterization for these sets is in general not trivial and increases the conservatism. In contrast, the approach proposed in this chapter is tight for arbitrary distributions and constraints given by (3.2a).

The following example illustrates this disadvantage of jointly optimizing nominal constraints which yields a given overall violation probability or similarly jointly determining the parameters for the confidence region of the disturbance. In particular, it demonstrates that jointly optimizing multiple parameters to determine offline a minimal size confidence region might lead to unexpectedly conservative results equivalent to a deterministic robust program.

Example 3.23. Let $x = z + e$ be a one dimensional random variable with $z = \mathbb{E}\{x\}$ and e having a non symmetric pdf, e.g., $e \sim \text{Gamma}(2, 0.1) - 0.2$. Let a joint chance constraint be given by $\mathbb{P}\{|x| \leq 1\} \geq 1 - \varepsilon$. Optimizing the bounds of nominal constraints $-\eta_2 \leq z \leq \eta_1$ jointly, e.g.,

$$\begin{aligned} \max_{\eta_1, \eta_2} \ & \eta_1 + \eta_2 \\ \text{s.t.} \ & \mathbb{P}\{\eta_1 \leq 1 - e,\ \eta_2 \leq 1 + e\} \geq 1 - \varepsilon \end{aligned}$$

or optimizing the bounds of a confidence interval $[\gamma_1, \gamma_2]$ for e

$$\begin{aligned} \max_{\gamma_1, \gamma_2} \ & \gamma_2 - \gamma_1 \\ \text{s.t.} \ & \mathbb{P}\{\gamma_1 \leq e \leq \gamma_2\} \geq 1 - \varepsilon \end{aligned}$$

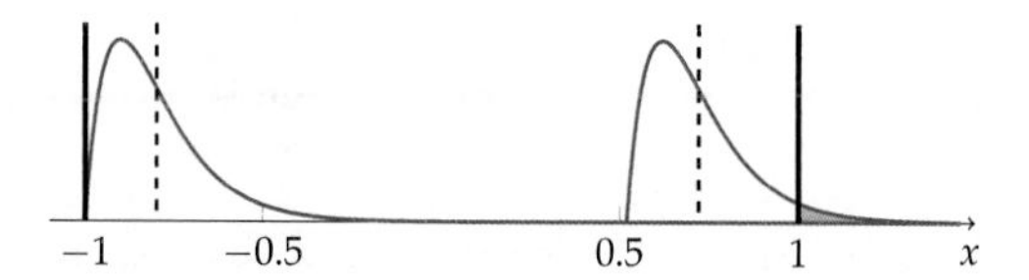

Figure 3.6. Constraints $|x| \leq 1$ (solid line), resulting tightened constraints for the nominal state z (dashed line) and probability density functions for x with $z = -0.8$ and $z = 0.73$, respectively (blue). A joint chance constraint evaluation as described in Example 3.23 leads to a biased outcome: The lower bound is satisfied with probability 1, whereas the upper bound is satisfied with probability $1 - \varepsilon$.

to derive the constraint $-1 - \gamma_1 \leq z \leq 1 - \gamma_2$ leads to a biased outcome. With $\varepsilon = 0.05$, the result is $\eta_1 \approx 0.73$, $\eta_2 = 0.8$ and $\gamma_1 = -0.2$, $\gamma_2 \approx 0.27$, respectively. As illustrated in Figure 3.6, the constraint $x \geq -1$ holds with zero probability of violation and $x \leq 1$ with ε probability of violation. If x is to be maximized, the result of the deterministic problem with nominal constraints will be equal to the solution of the original chance constrained problem. If x is to be minimized, the result will coincide with the solution of the robust problem.

Remark 3.24. If the probability density function of e is constant in some open set, e.g., uniform disturbance model, both optimizations in Example 3.23 might not have a unique optimizer. The solution and thereby the conservativeness of the nominal constraints then depends on the chosen optimization algorithm and initial value. This is a standard problem in determining a confidence region.

Remark 3.25. The described difficulties can be circumvented if the right hand side of the nominal constraints can be given as an explicit function of the violation probability ε_j. Using Boole's Inequality and dynamic risk allocation as proposed, e.g., in (Ono and Williams, 2008), the conservatism can be reduced at the cost of more complex online computations.

We conclude by emphasizing that the proposed approach of determining nominal constraints based on single chance constraints gives the best worst-case value in terms of conservativeness compared to directly approximating a joint chance constraint by means of nominal constraints.

Limitations of the proposed approach

On a conceptual level, the main limitation of the proposed algorithm is the restriction to random variables with bounded support. Yet this is to be expected if unstable system matrices and hard input constraints are included in the problem setup. An ad-hoc solution to deal with this limitation is to either use a penalty function reformulation or consider a confidence region with large probability mass instead of an outer bound. Based on standard results one can derive first exit times and the probability of the MPC optimization becoming infeasible within a given time-frame. However, since the proposed algorithm is less conservative, it can be employed with a comparatively large bounding set.

A related shortcoming is due to the choice of using chance constraints pointwise in time. In the current setup, the admissible set is limited to states where constraint satisfaction can, up to a chosen probability, be guaranteed within the next sampling time. In contrast, a further substantial relaxation could be achieved by restricting the frequency of constraint violations for sample paths over a finite or infinite time window.

From a computational point of view, the main limitation is the derivation of the robust control invariant set, in particular the set projections necessary to determine $\mathbb{X}_N^\infty$. While this is, at least for linear systems, a well understood topic with existing algorithms implemented in ready to use toolboxes, it nonetheless complicates the approach for applications with a "large" state space. However, similar complexity issues arise in standard tube MPC algorithms that are based on polytopic sets.

Summary

In this section, we have developed a simplified stochastic model predictive control algorithm for linear systems with additive disturbance modeled by a stochastic process. The algorithm unifies the results presented by Kouvaritakis et al. (2010) and Korda et al. (2011) allowing to balance convergence speed and performance guarantees against the size of the feasible region. Through explicitly considering the requirements of recursive feasibility and stability separately, we obtained a significant reduction in conservatism compared to state of the art SMPC algorithms that provide rigorous stability guarantees. Absolute bounds of the disturbance were used in a novel first step constraint to guarantee robust recursive feasibility and the stochastic description was exploited to prove stability. In particular, under mild assumptions, asymptotic stability with probability one of the set $\mathbb{X}_\infty$ has been proven, a result which is new in the stochastic MPC literature. From a computational point of view, instead of solving a stochastic optimal control problem in each iteration, the online optimization has been reduced to a linearly constrained quadratic program and is thus of comparable complexity as nominal MPC.

3.2 Linear systems with multiplicative disturbance

In the following, the computational approach is extended to linear systems with a multiplicative disturbance model. As before, the goal is to approximate the original stochastic program by a computationally tractable, deterministic optimization program which is solved online. We keep the approach of using relaxed constraints in combination with a first step constraint but directly approximate the resulting stochastic program as the separation principle cannot be applied.

3.2.1 Problem setup

Consider system dynamics (2.4) that are linear in the state and input, given by

$$x_{k+1} = A(w_k)x_k + B(w_k)u_k. \tag{3.22}$$

The system matrices $A : \mathbb{R}^{m_w} \to \mathbb{R}^{n\times n}$ and $B : \mathbb{R}^{m_w} \to \mathbb{R}^{n\times m}$ are, possibly nonlinear, continuous functions in the stochastic disturbance w_k satisfying the following assumption.

Assumption 3.4 (Disturbance sequence). *The disturbances $w_k \in \mathbb{R}^{m_w}$ for $k \in \mathbb{N}$ are realizations of independent and identically distributed random variables W_k with distribution $\mathbb{P}_w$ and support $\mathbb{W}$. Let $\mathbb{G} = \{(A(w), B(w))\}_{w\in\mathbb{W}}$. A polytopic outer approximation $\bar{\mathbb{G}} = \mathrm{co}\{(A^j, B^j)_{j\in\mathbb{N}_1^{N_c}}\} \supseteq \mathbb{G}$ exists and is known.*

Remark 3.26. As in the previous section, the outer approximation $\bar{\mathbb{G}}$ is necessary to derive a robust recursively feasible algorithm. Similar to the discussion in Section 3.1.5 it could be relaxed, e.g., to a confidence region, with the above stated consequences. Since the outer bound is used only for a one-step prediction, it does not lead to as conservative results as in robust MPC.

Given the state x_k at time k, the predicted, future states $x_{l|k}$ are modeled by a stochastic process satisfying

$$x_{l+1|k} = A(W_{l+k})x_{l|k} + B(W_{l+k})u_{l|k} \qquad x_{0|k} \overset{a.s.}{=} x_k. \tag{3.23}$$

The state and input constraint sets are assumed to be given by convex polytopes containing the origin, specifically $\mathbb{X} = \{x \in \mathbb{R}^n \mid Hx \leq 1\}$ and $\mathbb{U} = \{u \in \mathbb{R}^m \mid Gu \leq 1\}$ with $H \in \mathbb{R}^{p\times n}$ and $G \in \mathbb{R}^{q\times m}$. Similar to Section 3.1, we assume hard input constraints and the state constraints are formulated as individual chance constraints for each hyperplane, that is

$$\mathbb{P}\{[H]_j x_{1|k} \leq 1 \mid x_k\} \geq 1 - [\varepsilon]_j, \qquad \forall j \in \mathbb{N}_1^p \tag{3.24a}$$

$$Gu_k \leq 1 \tag{3.24b}$$

with $\varepsilon \in [0,1]^p$. As before, the running cost ℓ is assumed to be given by the quadratic function $\ell(x,u) = \|x\|_Q^2 + \|u\|_R^2$ with $Q \in \mathbb{R}^{n\times n}$, $Q \succ 0$, $R \in \mathbb{R}^{m\times m}$, $R \succ 0$.

Input parameterization and basic stability assumption

To derive a computationally tractable algorithm, we assume an input parameterization

$$u_{l|k}(x_{l|k}) = Kx_{l|k} + v_{l|k}, \tag{3.25}$$

with the feed forward input $v_{l|k} \in \mathbb{R}^m$ that is optimized online by the SMPC algorithm and a constant prestabilizing gain $K \in \mathbb{R}^{m\times n}$. Analogous to the basic stability assumption in Chapter 2, we make the following assumption on the prestabilizing gain K, the existence of a suitable terminal set and terminal cost $V_f(x) = \|x\|_P^2$.

Assumption 3.5 (Basic stability assumption, linear systems). *There exists a non-empty terminal constraint set $\mathbb{X}_f = \{x \mid H_f x \leq 1\}$, $H_f \in \mathbb{R}^{r\times n}$ that is robust forward invariant for* (3.22) *under the control law $u_k = Kx_k$. Furthermore, the resulting closed-loop state and input trajectories satisfy the constraints* (3.24). *There exists $P \in \mathbb{R}^{n\times n}$ such that*

$$Q + K^\top RK + \mathbb{E}\left[A_{cl}(W_k)^\top P A_{cl}(W_k)\right] - P \preceq 0 \tag{3.26}$$

with $A_{cl}(W_k) = A(W_k) + B(W_k)K$.

Note that if the matrices A and B are affine functions in the disturbance W_k and the random variables $\{[W_k]_j\}_{j\in\mathbb{N}_1^{m_w}}$ are independent, then (3.26) can be rewritten as an LMI to derive a control gain K and matrix P. Otherwise the outer bound $\bar{\mathbb{G}}$ can be used to derive a quadratically stabilizing control gain by standard robust control methods (Boyd et al., 1994).

3.2.2 SMPC design using finite sample approximations

Similar to the previous section, we relax the SMPC constraints to

$$\begin{aligned}
\mathbb{P}\{[H]_j x_{l|k} \leq 1 \mid x_k\} &\geq 1 - [\varepsilon]_j & \forall j \in \mathbb{N}_1^p,\ l \in \mathbb{N}_1^N \qquad &(3.27a)\\
\mathbb{P}\{[G]_j u_{l|k} \leq 1 \mid x_k\} &\geq 1 - [\varepsilon_u]_j & \forall j \in \mathbb{N}_1^q,\ l \in \mathbb{N}_0^{N-1}, \qquad &(3.27b)\\
\mathbb{P}\{[H_f]_j x_{N|k} \leq 1 \mid x_k\} &\geq 1 - [\varepsilon_f]_j & \forall j \in \mathbb{N}_1^r, \qquad &(3.27c)
\end{aligned}$$

with $\varepsilon_u, \varepsilon_f \in (0,1)$. Additionally, the initial state $x_{0|k}$ is kept as an artificial optimization variable. With this the constraint set can be split into a probabilistic constraint $\mathbf{v}_{N|k} \in \mathbb{D}^P$, where $\mathbb{D}^P$ is defined by chance constraints, and a deterministic constraint $\mathbf{v}_{N|k} \in \mathbb{D}^D(x_k)$, which is parameterized by x_k. The online optimization can then generically be written as a multi-parametric program

$$\begin{aligned} &\min_{\mathbf{v}_{N|k}} J_N(x_k, \mathbf{v}_{N|k}) \\ &\text{s.t. } \mathbf{v}_{N|k} \in \mathbb{D}^P \cap \mathbb{D}^D(x_k). \end{aligned} \tag{3.28}$$

The key observation is that only the deterministic constraint is parameterized by x_k and changes over time, while the set given by chance constraints $\mathbb{D}^P$ is constant. Since in general the chance constraints still make (3.28) an intractable problem, in the following we proceed by replacing $\mathbb{D}^P$ with a convex inner approximation $\mathbb{D}^S$. In particular, we will derive a polyhedral inner approximation by drawing offline a finite number of disturbance samples and employ results from statistical learning theory to give probabilistic guarantees on the resulting approximation.

A sampling subset result

Let $a : \Omega_a \to \mathbb{R}^{1\times d}$ be a real-valued, multivariate random variable with distribution $\mathbb{P}_a$ and define probabilities $\varepsilon, \delta \in (0,1)$. Let the set $\mathbb{D}^P$ be given by a single chance constraint

$$\mathbb{D}^P = \left\{ x \in \mathbb{R}^d \mid \mathbb{P}_a\{ax \leq 1\} \geq 1 - \varepsilon \right\}, \tag{3.29}$$

that is, by those x, where the constraint $ax \leq 1$ is satisfied with at least probability $1-\varepsilon$. Furthermore, let $a^{(i)}$, $i \in \mathbb{N}_1^{N_s}$ be N_s random samples of a and define the second set

$$\mathbb{D}^S = \left\{ x \in \mathbb{R}^d \mid a^{(i)}x \leq 1, \quad i \in \mathbb{N}_1^{N_s} \right\}. \tag{3.30}$$

Note that while $\mathbb{D}^P$ is a deterministic set, $\mathbb{D}^S$ is a random set, depending on the realizations $a^{(i)}$.

Obviously, for $N_s \to \infty$ points in $\mathbb{D}^S$ satisfy the chance constraint with probability 1. With the following proposition we derive an explicit sample complexity $\mathbb{N}_s$ such that, with a given confidence δ, the set $\mathbb{D}^S$, described by randomly sampled constraints, is a subset of the set $\mathbb{D}^P$, described by the original chance constraint. The confidence is measured with the product probability $\mathbb{P}_a^{N_s}$ according to which the multisample $(a^{(i)})_{i\in\mathbb{N}_1^{N_s}}$ is drawn.

Proposition 3.27 (Sampling approximation of chance constraint sets). *For any probability $\varepsilon \in (0,1)$, confidence $\delta \in (0,1)$, and sample complexity $N_s \in \mathbb{N}$ with*

$$N_s \geq \frac{5}{\varepsilon}\left(\ln\frac{4}{\delta} + d\ln\frac{40}{\varepsilon}\right) \tag{3.31}$$

it holds

$$\mathbb{P}_a^{N_s}\left\{\mathbb{D}^S \subseteq \mathbb{D}^P\right\} \geq 1-\delta. \tag{3.32}$$

If $\varepsilon \in (0, 0.14)$, the result holds true for the improved bound

$$N_s \geq \frac{4.1}{\varepsilon}\left(\ln \frac{21.64}{\delta} + 4.39 d \log_2\left(\frac{8e}{\varepsilon}\right)\right). \tag{3.33}$$

The proposition is based on statistical learning theory by proving an upper bound on the VC Dimension of the class of linear half spaces $\{x \in \mathbb{R}^d \mid ax \leq 1\}_{a \in \mathbb{R}^{1\times d}}$. With this, the explicit sample complexity is derived from Corollary 4 and Theorem 8 in (Alamo et al., 2009). The proof of the VC Dimension is given in Appendix B.2.

Remark 3.28. It is worth emphasizing that the sample complexity given in (3.31) and (3.33) is of the order $\frac{1}{\varepsilon}\ln\frac{1}{\varepsilon}$ and thus significantly larger than the bound given in (Calafiore and Fagiano, 2013a), which is based on scenario theory (Campi and Garatti, 2008; Calafiore, 2010) and of the order $\frac{1}{\varepsilon}$. However, this is to be expected as the requirement of the sampled set being a subset of the set given by chance constraints is much stronger than the requirement that the optimal solution of an optimization program subject to chance constraints satisfies these. The following example highlights this difference between offline and online sampling approximations applied to chance constraints in stochastic MPC.

Example 3.29 (Offline and online sampling approximation in SMPC)**.** Consider the following linear system with parameter $\gamma \in (0, 1)$

$$x_{k+1} = \begin{bmatrix} 1 & \gamma \\ -\gamma & 1 \end{bmatrix} x_k + \begin{bmatrix} 1 & 0 \\ 0 & 1 \end{bmatrix} u_k$$

subject to a single chance constraint

$$\mathbb{P}\left\{\begin{bmatrix}\cos(W_k) & \sin(W_k)\end{bmatrix} u_k \leq 1\right\} \geq 1-\varepsilon \tag{3.34}$$

with W_k uniformly distributed on the interval $[0, 2\pi]$.

The chance constraint (3.34) can be solved analytically and for $\varepsilon < 0.5$ the admissible set is equivalent to the set $\{u_k \in \mathbb{R}^2 \mid \|u_k\| \leq \frac{1}{\cos(\varepsilon\pi)}\}$. Figure 3.7 depicts the chance constraint along with a realization of sampled constraints: The admissible region of the chance constraint is given by the area inside the solid circle and the sampled hyperplanes are given by the solid lines.

For a quadratic running cost given by $\ell(x_k, u_k) = \|x_k\|^2 + \|u_k\|^2$ and $\|x_k\|$ sufficiently large, the optimal input is at a vertex of the polytope resulting from the sampled constraints. If new samples are drawn at each time step, the probability

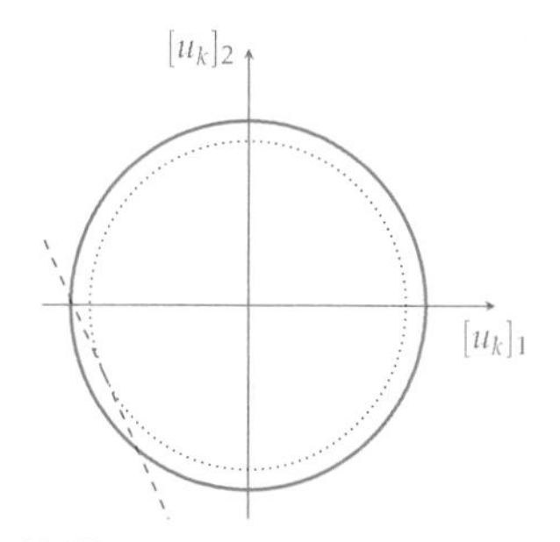

(a) Chance constraint (3.34) (solid circle) and one realization of the constraint (dashed line).

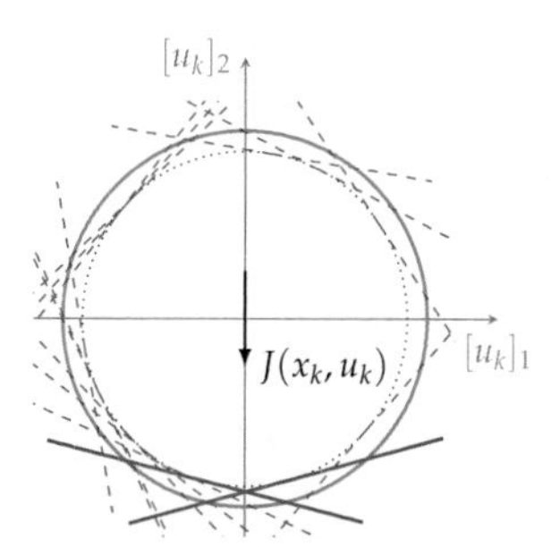

(b) Sampled constraints (dashed and solid lines) and active constraints (solid lines) for given objective function J.

Figure 3.7. Chance constraint (3.34) (solid circle), the corresponding robust constraint (dotted circle), and realizations of sampled constraints (lines). For a given objective function, the optimal solution satisfies the chance constraint if the intersection of the active constraints is within the solid circle. For (3.32) to hold, *all* vertices of the resulting polytope need to be within the solid circle.

of the chance constraint (3.34) being satisfied is equivalent to the probability of one *random* vertex being inside the solid circle. In contrast, when the same set of samples is used in each time step, then the optimal solution switches *deterministically* between the vertices, which leads to the stronger subset condition (3.32). In particular, for small values of γ and large $\|x_0\|$, the optimal solution can be 'trapped' at one vertex over multiple sampling times.

Note that the system and constraint can be easily transformed into the form (3.22) and (3.24a) by introducing a new state $\tilde{x}_k$ with dynamics $\tilde{x}_{k+1} = [\cos(\alpha)\ \sin(\alpha)]u_k$ and chance constraint $\mathbb{P}\{\tilde{x}_{l|k} \leq 1 \mid x_k\} \geq 1 - \varepsilon$.

Remark 3.30. A result similar to Proposition 3.27 but with discarded samples can be derived. Although this leads to a less conservative approximation of the chance constraint set, in the following this is not considered, since it would lead to solving a mixed integer program online and the constraint removal proposed in the next section is not easily possible.

Offline design of constraint set and cost functional

In order to exploit Proposition 3.27 for deriving a stochastic MPC design based on offline uncertainty sampling, we explicitly solve the system dynamics (3.23) with prestabilizing input (3.25) for the predicted states $\mathbf{x}_{N|k}$ and feed-forward input trajectory $\mathbf{v}_{N|k}$. With $\mathbf{W}_{N|k} = (W_{k+l})_{l\in\mathbb{N}_0^{N-1}}$ and suitable matrices $\Phi^0_{l|k}(\mathbf{W}_{N|k})$, $\Phi^u_{l|k}(\mathbf{W}_{N|k})$, and Γ_l which, for completeness, are given in Appendix C.1, we get

$$\begin{aligned} x_{l|k}(\mathbf{W}_{N|k}) &= \Phi^0_{l|k}(\mathbf{W}_{N|k})x_k + \Phi^u_{l|k}(\mathbf{W}_{N|k})\mathbf{v}_{N|k} \\ u_{l|k}(\mathbf{W}_{N|k}) &= Kx_{l|k} + v_{l|k} \\ &= K\Phi^0_{l|k}(\mathbf{W}_{N|k})x_k + (K\Phi^u_{l|k}(\mathbf{W}_{N|k}) + \Gamma_l)\mathbf{v}_{N|k}. \end{aligned} \tag{3.35}$$

Using the subset result developed above and the explicit solution (3.35), an inner approximation for the chance constraints (3.27) can be derived based on independent samples of the disturbance. For each $j \in \mathbb{N}_1^p$ and $l \in \mathbb{N}_1^N$ the chance constraints (3.27a) are equivalent to $(x_k, \mathbf{v}_{N|k}) \in \mathbb{X}_l^{P,j}$ with

$$\mathbb{X}_l^{P,j} = \left\{(x_k, \mathbf{v}_{N|k}) \in \mathbb{R}^n \times \mathbb{R}^{Nm} \mid \mathbb{P}\{[H]_j x_{l|k}(\mathbf{W}_{N|k}) \le 1\} \ge 1 - \varepsilon_j\right\}. \tag{3.36}$$

Let $\delta \in (0,1)$ be a user defined confidence level and for each l let the sample complexity $N_l \in \mathbb{N}$ be chosen according to Proposition 3.27 with $d = n + lm$. For each $j \in \mathbb{N}_1^p$ let $\{\mathbf{w}_j^{(i)}\}_{i\in\mathbb{N}_1^{N_l}}$ be a set of N_l independently drawn samples from $\mathbf{W}_{N|k}$. The finite sample approximation of (3.36) is given by

$$\mathbb{X}_l^{S,j} = \left\{(x_k, \mathbf{v}_{N|k}) \in \mathbb{R}^n \times \mathbb{R}^{Nm} \mid [H]_j x_{l|k}(\mathbf{w}_j^{(i)}) \le 1, \quad \forall i \in \mathbb{N}_1^{N_l}\right\}. \tag{3.37}$$

According to Proposition 3.27, we have $\mathbb{X}_l^{S,j} \subseteq \mathbb{X}_l^{P,j}$ with confidence $1-\delta$ and hence the relaxed chance constraint (3.27a) is satisfied.

Similarly, the relaxed input and terminal constraints (3.27b), (3.27c) can be approximated. Let the sample complexities $N_l^u, N_f \in \mathbb{N}$ be chosen according to Proposition 3.27 with $d = n + (l+1)m$ and $d = n + Nm$, respectively. For each $j_u \in \mathbb{N}_1^q$, $j_f \in \mathbb{N}_1^r$, let $\{\mathbf{w}_{u,j_u}^{(i)}\}_{i\in\mathbb{N}_1^{N_l^u}}, \{\mathbf{w}_{f,j_f}^{(i)}\}_{i\in\mathbb{N}_1^{N_f}}$ be sets of N_l^u respectively N_f independently drawn samples from $\mathbf{W}_{N|k}$. The finite sample approximation of (3.27b) and (3.27c) are given by

$$\begin{aligned} \mathbb{U}_l^{S,j_u} &= \left\{(x_k, \mathbf{v}_{N|k}) \in \mathbb{R}^n \times \mathbb{R}^{Nm} \mid [G]_{j_u} u_{l|k}(\mathbf{w}_{u,j_u}^{(i)}) \le 1, \quad \forall i \in \mathbb{N}_1^{N_l^u}\right\}, \\ \mathbb{X}_f^{S,j_f} &= \left\{(x_k, \mathbf{v}_{N|k}) \in \mathbb{R}^n \times \mathbb{R}^{Nm} \mid [H_f]_{j_f} x_{N|k}(\mathbf{w}_{f,j_f}^{(i)}) \le 1, \quad \forall i \in \mathbb{N}_1^{N_f}\right\} \end{aligned} \tag{3.38}$$

Note that the sets $\mathbb{X}_l^{S,j}$, $\mathbb{U}_l^{S,j}$, and $\mathbb{X}_f^{S,j}$ are convex polytopes given by linear constraints, which, due to the random sampling, are in general highly redundant. Since the samples are drawn offline, redundant constraints can be easily removed which decreases the complexity of the online optimization.

Remark 3.31. Redundant constraints can be easily identified by solving a series of linear optimization programs as shown in the following basic algorithm. This is a standard problem with efficient implementations available, e.g., (Herceg et al., 2013). The constraints could be further simplified by any inner approximation, which has been done in the example in Section 3.2.4.

Algorithm 3.32 (Basic constraint removal).
Require: Constraint $Hx \leq h$
Ensure: $H \in \mathbb{R}^{n_c \times n}$, $h \in \mathbb{R}^{n_c}$
while $i \leq n_c$ **do**
solve

$$h_i^* = \max_x \; [H]_i x \quad \text{s.t. } [H]_k x \leq [h]_k \quad \forall k \in \mathbb{N}_1^{n_c} \setminus i$$

if $h_i^* \leq [h]_i$ **then**
$H \leftarrow H \setminus [H]_i$
$h \leftarrow h \setminus [h]_i$
$n_c = n_c - 1$
else
i=i+1
end if
end while
return H, h

In the following, we assume that the intersection of the sampled constraint sets (3.37) and (3.38), possibly with redundant constraints removed, is given by

$$\mathbb{D}^S = \left\{ (x_k, \mathbf{v}_{N|k}) \in \mathbb{R}^n \times \mathbb{R}^{Nm} \mid \begin{bmatrix} \tilde{H} & H \end{bmatrix} \begin{bmatrix} x_k \\ \mathbf{v}_{N|k} \end{bmatrix} \leq h \right\}. \tag{3.39}$$

Cost functional Given the input parameterization (3.25) and explicit solution (3.35), the expected value of the SMPC cost functional can be solved explicitly offline. This leads to the deterministic, convex, quadratic cost function

$$J_N(x_k, \mathbf{v}_{N|k}) = \begin{bmatrix} x_k^\top & \mathbf{v}_{N|k}^\top \end{bmatrix} \tilde{Q} \begin{bmatrix} x_k \\ \mathbf{v}_{N|k} \end{bmatrix} \tag{3.40}$$

with appropriate $\tilde{Q}$ given in Appendix C.1.

Recursive feasibility

As discussed in Section 3.1.2, a stochastic MPC algorithm with the considered constraint relaxation is not recursively feasible. In the following, we briefly present the derivation of the additional first step constraint, proposed in Section 3.1.2, adapted to the current setup of linear systems with multiplicative disturbance.

The N-step set and feasible first input is given by

$$\tilde{\mathbb{X}}_N^u = \left\{ (x_k, v_{0|k}) \in \mathbb{R}^n \times \mathbb{R}^m \;\middle|\; \begin{array}{l} \exists v_{1|k}, \dots, v_{N-1|k} \in \mathbb{R}^m \\ \text{satisfying } (x_k, \mathbf{v}_{N|k}) \in \mathbb{D}^S \end{array} \right\}$$

The set $\tilde{\mathbb{X}}_N^u$ defines the feasible states and first inputs of the sampling based approximation to the stochastic MPC finite horizon program. For the first step constraint, let

$$\mathbb{X}_N^\infty = \{x \in \mathbb{R}^n \mid H_\infty x \leq h_\infty\}$$

be a robust control invariant polytope for the system (3.22) with $(A(W_k), B(W_k)) \in \mathbb{G}$ and constraint $(x, u) \in \tilde{\mathbb{X}}_N^u$. Finally, the additional constraint set to render the SMPC optimization recursively feasible is given by

$$\mathbb{D}^R = \left\{ (x_k, \mathbf{v}_{N|k}) \in \mathbb{R}^n \times \mathbb{R}^{Nm} \mid H_\infty A_{cl}^j x_k + H_\infty B^j v_{0|k} \leq h_\infty,\ \forall j \in \mathbb{N}_1^{N_c} \right\} \tag{3.41}$$

with $(A^j,\ B^j)$ being the vertices of the polytopic outer approximation $\bar{\mathbb{G}}$ and $A_{cl}^j = A^j + B^j K$.

Remark 3.33. Given the polytopic outer bound $\bar{\mathbb{G}}$, the robust control invariant set $\mathbb{X}_N^\infty$ can be computed via recursions as discussed in Section 3.1.2. Further details on algorithms for the considered setup can be found in, e.g., Section 5.3 of (Blanchini and Miani, 2015).

3.2.3 Stochastic MPC algorithm and closed-loop properties

The complete stochastic MPC algorithm can be divided into two parts. First, an offline design phase which includes the cost evaluation, disturbance sampling and invariant set computation. Second, the online MPC algorithm which has been reduced to solving in each iteration a deterministic, convex linearly constrained quadratic program. In the following, we summarize the complete algorithm and analyze its closed-loop properties.

Algorithm 3.34 (Stochastic MPC for linear systems with multiplicative disturbance).

Offline: Evaluate the expected value (C.1) to determine the explicit cost matrix $\tilde{Q}$ of the nominal cost (3.40). Draw a sufficiently large number of samples to determine the sampled constraints (3.37) and (3.38). Remove redundant constraints to get (3.39) and determine the first step constraint (3.41).
Online: For each time step $k = 0, 1, 2, \dots$

1. Measure the current state x_k.
2. Determine the minimizer $\mathbf{v}^*_{N|k}$ of the quadratic cost (3.40) subject to the linear constraints (3.39) and (3.41)

$$\begin{aligned} \mathbf{v}^*_{N|k} = \arg\min_{\mathbf{v}_{N|k}} & \begin{bmatrix} x_k^\top & \mathbf{v}_{N|k}^\top \end{bmatrix} \tilde{Q} \begin{bmatrix} x_k \\ \mathbf{v}_{N|k} \end{bmatrix} \\ \text{s.t.} \ & (x_k, \mathbf{v}_{N|k}) \in \mathbb{D}^S \cap \mathbb{D}^R. \end{aligned} \tag{3.42}$$

3. Apply the SMPC feedback law $\kappa(x_k) = Kx_k + v^*_{0|k}$.

As before, the set of admissible input sequences of the SMPC optimization (3.42) for a given state x_k is denoted by

$$\mathbb{D}_N(x_k) = \{\mathbf{v}_{N|k} \in \mathbb{R}^{mN} \mid (x_k, \mathbf{v}_{N|k}) \in \mathbb{D}^S \cap \mathbb{D}^R\},$$

the set of feasible initial conditions by $\mathbb{X}_N = \{x \in \mathbb{R}^n \mid \mathbb{D}_N(x) \neq \emptyset\}$, and the optimal value function by V_N.

Remark 3.35. To further simplify the computation, the redundant constraints obtained combining (3.39) and (3.41) should be removed offline, as well. In the algorithm above they are kept separate to emphasize the conceptually different constraints.

Properties of the proposed stochastic MPC scheme

Analogous to the results presented in Section 3.1, recursive feasibility of the SMPC optimization and constraint satisfaction in closed loop holds due to the proposed offline design. Yet, as the stronger property of feasibility of the candidate solution is not satisfied robustly, we need an explicit upper and lower bound on the optimal value function to prove asymptotic stability.

Assumption 3.6 (Bound on V_N). *There exist symmetric matrices $P_l, P_u \in \mathbb{R}^{n\times n}$, $P_l \succ 0$, $P_u \succ 0$ such that $\|x\|^2_{P_l} \leq V_N(x) \leq \|x\|^2_{P_u}$ holds $\forall x \in \mathbb{X}_N$.*

Similar to the Lipschitz bound in Section 3.1.3, the matrices P_l and P_u allow to derive bounds on the cost increase if the candidate solution does not remain feasible. The matrix P_l is naturally given by the unconstrained infinite horizon cost matrix P. Since the optimal value function V_N is known to be convex and piecewise quadratic, the upper bound can be determined based on the vertices of the feasible set $\mathbb{X}_N$.

Finally, the closed-loop properties are summarized in the following theorem.

Theorem 3.36 (Sampling based SMPC closed-loop properties). *Consider the sampling based stochastic MPC Algorithm 3.34 in closed loop with the linear system 3.22 and suppose Assumptions 3.4 as well as 3.5 are satisfied. The SMPC optimization 3.42 is robust recursively feasible, that is $x_k \in \mathbb{X}_N$ implies $A(w_k)x_k + B(w_k)\kappa(x_k) \in \mathbb{X}_N$ for every realization $w_k \in \mathbb{W}$. For $x_0 \in \mathbb{X}_N$, the state and input trajectories satisfy the hard input constraints* (3.24b) *robustly and the probabilistic state constraints* (3.24a) *for each $j \in \mathbb{N}_1^p$ with confidence $1-\delta$.*

Suppose additionally Assumption 3.6 is satisfied and for each $k \in \mathbb{N}$ the candidate solution $\tilde{v}_{l|k+1} = v^*_{l+1|k}$, $\tilde{v}_{N-1|k+1} = 0$ remains feasible with at least probability $1-\varepsilon_{\bar{f}} > 0$. If for all $(A,B) \in \mathbb{G}$

$$\begin{bmatrix} Q - \frac{\varepsilon_{\bar{f}}}{1-\varepsilon_{\bar{f}}}(A^\top P_u A - P_l) & -\frac{\varepsilon_{\bar{f}}}{1-\varepsilon_{\bar{f}}}A^\top P_u B \\ -\frac{\varepsilon_{\bar{f}}}{1-\varepsilon_{\bar{f}}}B^\top P_u A & R - \frac{\varepsilon_{\bar{f}}}{1-\varepsilon_{\bar{f}}}(B^\top P_u B) \end{bmatrix} \succ 0, \tag{3.43}$$

then the origin is asymptotically stable with probability 1 under the proposed SMPC scheme.

Proof. By definition of $\mathbb{D}^R$, $(x_k, \mathbf{v}_{N|k}) \in \mathbb{D}^R$ implies $x_{k+1} \in \mathbb{X}_N^\infty$ for all $w_k \in \mathbb{W}$. Since $\mathbb{X}_N^\infty \subseteq \mathbb{X}_N$ recursive feasibility follows.

Hard input constraint satisfaction follows from robust recursive feasibility since the input constraint $Gu_{0|k} \leq 1$ does not rely on sampling. For each $j \in \mathbb{N}_1^p$, $x_k \in \mathbb{X}_N$ we have $(x_k, \mathbf{v}^*_{N|k}) \in \mathbb{X}_1^{S,j}$ and by Proposition 3.27 it holds $\mathbb{X}_1^{S,j} \subseteq \mathbb{X}_1^{P,j}$ with confidence $1-\delta$. Hence, for all $x_k \in \mathbb{X}_N$ the chance constraint $\mathbb{P}\{[H]_j x_{1|k} \leq 1 \mid x_k\} \geq 1-\varepsilon$ is satisfied with confidence $(1-\delta)$.

The proof of asymptotic stability follows by using V_N as stochastic Lyapunov function and, analogously to the proof of Theorem 3.9, explicitly taking into account possible infeasibility of the candidate solution. We first consider the case that the candidate solution is feasible at time $k+1$. Let $\mathbb{E}\{V_N(x_{k+1}) \mid x_k, \tilde{\mathbf{v}}_{N|k+1} \text{ admissible}\}$ be the expected optimal value at time $k+1$, conditioning on the state at time k and feasibility of the candidate solution $\tilde{\mathbf{v}}_{N|k+1}$.

$$\begin{aligned}
&\mathbb{E}\left\{V_N(x_{k+1}) \mid x_k, \tilde{\mathbf{v}}_{N|k+1} \text{ admissible}\right\} - V_N(x_k) \\
&\leq \mathbb{E}\left\{\sum_{l=0}^{N-1}\left(\|\tilde{x}_{l|k+1}\|_Q^2 + \|\tilde{u}_{l|k+1}\|_R^2\right) + \|\tilde{x}_{N|k+1}\|_P^2 \mid x_k\right\} \\
&\quad - \mathbb{E}\left\{\sum_{l=0}^{N-1}\left(\|x^*_{l|k}\|_Q^2 + \|u^*_{l|k}\|_R^2\right) + \|x^*_{N|k}\|_P^2 \mid x_k\right\} \\
&= \mathbb{E}\left\{\|x^*_{N|k}\|^2_{(Q+K^\top RK)} + \|A_{cl}(W_{k+N})x^*_{N|k}\|_P^2\right. \\
&\qquad \left. -\|x^*_{N|k}\|_P^2 \mid x_k\right\} - \|x^*_{0|k}\|_Q^2 - \|u^*_{0|k}\|_R^2 \\
&\leq -\|x_k\|_Q^2 - \|u_k\|_R^2.
\end{aligned}$$

In case the candidate solution does not remain feasible, Assumption 3.6 can be employed to compute an upper bound of the Lyapunov function increase

$$\mathbb{E}\left\{V_N(x_{k+1}) \mid x_k, \tilde{\mathbf{v}}_{N|k+1} \text{ not feasible}\right\} - V_N(x_k) \leq \max_{(A,B)\in\mathbb{G}} \|Ax_k + Bu_k\|^2_{P_u} - \|x_k\|^2_{P_l}.$$

Let $\lambda_{\min}$ be a lower bound of the smallest eigenvalue of the matrix (3.43) for all $(A, B) \in \mathbb{G}$. Then, by the law of total probability, it holds

$$\begin{aligned}
&\mathbb{E}\left\{V_N(x_{k+1}) \mid x_k\right\} - V_N(x_k) \\
&\leq -(1-\varepsilon_f)\left(\|x_k\|_Q^2 + \|u_k\|_R^2\right) + \varepsilon_f\left(\max_{(A,B)\in\mathbb{G}} \|Ax_k + Bu_k\|^2_{P_u} - \|x_k\|^2_{P_l}\right) \\
&\leq -(1-\varepsilon_f)\lambda_{\min}\|x_k\|_2^2,
\end{aligned}$$

which is a sufficient condition for asymptotic stability with probability 1. ■

Remark 3.37. It suffices to check (3.43) for the vertices of the set $\bar{\mathbb{G}}$, since it can be recast as an LMI in A and B using the Schur complement. If $\bar{\mathbb{G}}$ is given by interval matrices, the results in (Alamo et al., 2008) can be applied to further reduce the number of LMIs that need to be checked.

Remark 3.38. Note that, due to sampling offline, the confidence $1-\delta$ remains the same for all times $k \geq 1$. This determines the probability that chance constraint satisfaction does not hold and should therefore be chosen sufficiently high. The confidence holds for each chance constraint individually and by independence of the events $\{\mathbb{X}_1^{S,j} \subseteq \mathbb{X}_1^{P,j}\}_{j\in\mathbb{N}_1^p}$ all chance constraints are satisfied simultaneously with at least confidence $(1-\delta)^p$.

3.2.4 Numerical example

To illustrate the results, we briefly present the application of the presented SMPC algorithm in a numerical example. Consider a linear system given by (3.22) with

$$A(w_k) = \begin{bmatrix} 1 + w_{k,1} & 0.0075\left(w_{k,2} + \frac{1}{w_{k,3}^2}\right) \\ -0.14 w_{k,2} & w_{k,1} + \frac{1}{w_{k,3}^2} \end{bmatrix}, \quad B(w_k) = \begin{bmatrix} 4.8 \\ 0.1 w_{k,2} \end{bmatrix},$$

where $w_k = [w_{k,1}, w_{k,2}, w_{k,3}]$ are iid random variables with $w_{k,1}$ uniformly distributed on the interval $[-0.1, 0.1]$ and $w_{k,2}, w_{k,3}$ uniformly distributed on the interval $[0.9, 1.1]$. The system is subject to hard input constraints $|u_k| \leq 0.5$ as well as chance constraints on the state $|[x_k]_1| \leq 2$, $|[x_k]_2| \leq 3$ that should be satisfied with probability $1 - \varepsilon = 0.95$ each.

In the simulations, the MPC cost weights were chosen to be $Q = \text{diag}(1, 10)$, $R = 1$ and the prediction horizon to be $N = 8$. For disturbance attenuation in the predictions, the input was parametrized with the prestabilizing gain $K = [-0.49\ 1.86]$. The confidence level for the sampling based inner approximation has been set to $1 - \delta = 1 - 10^{-6}$. To reduce the number of constraints in the online optimization, redundant constraints in $\mathbb{D}^S \cap \mathbb{D}^R$ have been removed. To this end, the sampled constraints have been clustered and the constraint removal has been performed recursively leading to more optimizations being solved but each with a reduced size. Furthermore, a simple inner approximation $\widetilde{\mathbb{D}}^S$ of $\mathbb{D}^S$ has been employed, which led to a further significant reduction in the number of constraints. With $0.99\mathbb{D}^s \subseteq \widetilde{\mathbb{D}}^s \subseteq \mathbb{D}^S$ the number of constraints could be reduced by 97.6% from initially 854 514 sampled constraints to 20 187. With $0.98\mathbb{D}^s \subseteq \widetilde{\mathbb{D}}^s \subseteq \mathbb{D}^S$ the number of constraints could be reduced by 99.2% down to 6 650.

The simulations were performed with Matlab using the standard interior point algorithm of `quadprog`. Out of 1 000 simulations with random initial condition, the median reported computation time to solve the online optimization (3.42) with the above discussed inner approximation of 2% but no further preconditioning was 0.18s with a maximum of 0.23s on an Intel Core i7 with 3.4GHz. For comparison, in (Lorenzen et al., 2015a) the control problem has been solved using online sampling and Scenario Theory which required, including sample generation, an average computation time of 2.25s with a maximum of 4.15s.

Figure 3.8 shows the admissible region of an SMPC based on only the sampled constraints $\mathbb{D}^S$ jointly with the admissible region of the proposed SMPC with first step constraint. The difference in size can be interpreted as the price for robustness. In the red region, recursive feasibility of the SMPC optimization cannot be guaranteed. In the example, the difference could be decreased further by using a tighter robust bounding set $\bar{\mathbb{G}}$.

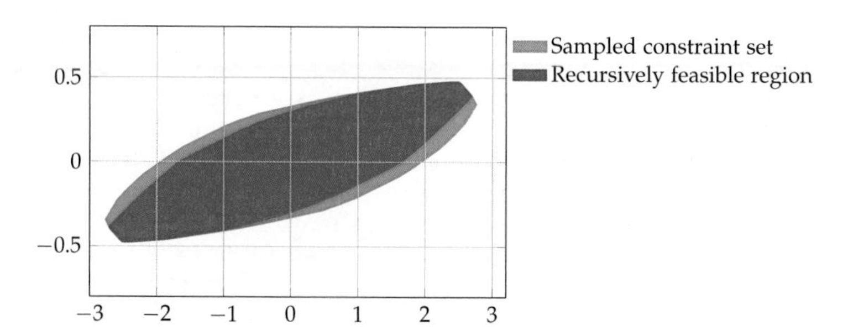

Figure 3.8. Projection of the set $\mathbb{D}^S$ and $\mathbb{D}^S \cap \mathbb{D}^R$ onto the state coordinates. Due to the additional constraint, the admissible set of the Stochastic MPC is smaller, but the online optimization is guaranteed to remain feasible.

3.2.5 Discussion and summary

To conclude this section on offline sampling based MPC for linear systems with multiplicative disturbance, we briefly discuss the advantages and limitations of the proposed approach. In particular, we link the algorithm to online sampling using scenario theory.

Link to scenario MPC

The proposed stochastic MPC is closely related to so-called scenario MPC (Calafiore and Fagiano, 2013b; Schildbach et al., 2014) where the disturbance sequence is sampled *online*. In each iteration, new samples are drawn to create the sampled constraint set $\mathbb{D}^S$, which are used without a first step constraint in the SMPC optimization 3.42.

The main advantage of the scenario MPC is an easy implementation of the algorithm without the tedious offline design. Furthermore, the sample complexity is based on scenario theory (Campi and Garatti, 2008; Calafiore, 2010; Tempo et al., 2013) rather than more classical results in statistical learning theory. Compared to the results in Proposition 3.27, far less samples are necessary in each optimization as discussed in Remark 3.28 and Example 3.29. Yet, new samples need to be drawn in each iteration which ultimately leads to the requirement of an infinite number of samples being available. The difference in the number of samples is based on the difference that scenario theory guarantees, with confidence $1 - \delta$, satisfaction of the original chance constraints for the *optimal solution* of the sampled program. In contrast, Proposition 3.27 implies that, with confidence $1 - \delta$, *all* admissible points of the sampled constraints $\mathbb{D}^S$ satisfy the original chance constraints.

The main disadvantage of scenario MPC is that recursive feasibility cannot be proven and hence neither constraint satisfaction nor asymptotic stability of the origin (Deori et al., 2014; Schildbach et al., 2014). In contrast, sampling the uncertainty offline allows to suitably augment the constraints in such a way that recursive feasibility can be guaranteed independent of the sample realization.

For a more detailed analysis of the difference between offline and online sampling in scenario based approximations to SMPC the reader is referred to (Lorenzen et al., 2017c).

Limitations of the proposed approach

On a conceptual level, the main limitations of the proposed approach are the restriction to iid disturbance sequences as well as the assumption of a known outer bound of the set $\mathbb{G}$. The former is a general limitation of SMPC algorithms based on offline approximation of the chance constrained program. A common solution is to augment the system by a filter which is used to generate colored noise based on an iid input sequence. For the latter, the same remedies as proposed in the previous section apply.

From a computational point of view, the main limitation is the sample complexity N_s, which is necessary to prove constraint satisfaction. The large number of sampled constraints significantly complicates the offline computations, in particular the set projection to compute $\mathbb{D}^R$. While N_s grows only linearly with the number of decision variables, roughly input dimension times horizon length, it grows with $\frac{1}{\varepsilon}\ln(\frac{1}{\varepsilon})$ with respect to the probabilistic level of constraint satisfaction ε. Yet, we highlight that samples need not be generated online and typically a significant part of the sampled constraints are redundant and can be removed. Nonetheless, for small values of ε suitable methods for constraint pruning that result in an inner approximation of the sampled set need to be employed. To overcome the difficulty involved in the orthogonal projection of polytopes, the admissible set could be approximated as proposed in (Scibilia et al., 2011). Alternatively, based on $\mathbb{D}^S$, a direct computation of invariant sets with a priory specified complexity could be performed similar to the algorithms presented in (Athanasopoulos et al., 2014). While this creates additional conservatism, it results in an efficiently solvable online optimization.

Finally, we remark the importance and advantage of having reduced the main problem to a fundamental question of general interest: Deriving a suitable inner approximation of a chance constraint set. Further results that can be applied instead of Proposition 3.27 and which simplify the proposed SMPC design, in particular lower bounds on the sample complexity, are to be expected in the future.

First results in this direction have been presented in (Alamo et al., 2018).

Summary

In this section, we have developed a computational control method to stochastic MPC for linear systems with multiplicative disturbance. The proposed algorithm is based on disturbance realizations, which are used in the offline calculations. Thereby, the approach is suitable for a data-driven design based on a finite number of samples.

We have shown that the original SMPC constraint can be split into a time-varying deterministic constraint and a set of chance constraints that are constant. This allowed to reduce the problem of offline sampling in MPC to the basic question of finding a lower bound on the sample complexity such that the sample approximation is a subset of the original chance constraint set. This question has been answered using results from statistical learning theory, finally leading to provable guarantees for constraint satisfaction in closed loop. The proposed approach allowed to introduce an additional constraint to guarantee robust recursive feasibility and, under further assumptions, asymptotic stability with probability 1 of the origin.

The results directly extend to nonlinear systems and constraints, provided an upper bound on the VC-dimension can be established.

3.3 Summary

The purpose of this chapter was to derive rigorous computational approaches to stochastic MPC for linear systems based on the conceptual algorithm presented in Chapter 2. Thereby, several contributions in the field of stochastic MPC in general and computationally tractable algorithms for linear systems in particular were obtained.

We separated the requirements of recursive feasibility and stability, which led to a significant reduction in conservatism of both algorithms. In particular, we highlighted the difference between existence of a solution and feasibility of an a priori known candidate solution. The relevance of the difference is unique to stochastic MPC. In the first part, this eventually led to an SMPC algorithm that unified previous results in the literature. In the second part, this enabled the use of a sampling based approximation, which still allowed to prove recursive feasibility of the online optimization and stability of the closed loop. The general theme was to introduce an additional constraint on the applied input, which is necessary and sufficient for recursive feasibility if the given disturbance bounds are tight and the set computations exact.

In both parts we presented computationally efficient sampling based approaches to approximate the chance constrained programs. In the second part we introduced a novel design based on sampling the disturbance offline, which is in line with the recommendations given in (Mayne, 2014). In particular, we formulated the requirements for constraint design as the fundamental question of finding an inner approximation of a set given by chance constraints and presented results based on statistical learning theory.

The results in this chapter can serve as starting point for different interesting research topics. The first step constraint can be exploited for further computational approaches to stochastic MPC. The question of finding an inner approximation, although not being trivial, deserves further research as it is of general interest. For a detailed discussion on these issues, the reader is referred to Section 6.2.

Chapter 4

Uncertainty and online model adaption

In the previous two chapters, we considered constrained control of systems under rapidly changing disturbances. The analysis in Chapter 2 eventually led to computationally tractable, local approximations for linear systems. The two critical assumptions, which enabled the presented approximations, were the disturbances being independent and identically distributed as well as direct measurement of the state. This led to the mathematical framework being that of a fully observable Markov decision process where the control policy could be derived through a local approximation around the measured state.

If the disturbances are correlated, e.g., unknown but constant or slowly changing model parameters, the state needs to be suitably augmented to recover a Markov chain again. Yet, since these additional states cannot be measured, the problem is that of a partially observable Markov decision process. In particular, as emphasized in Remark 2.8 the initial condition $x_{0|k}$ in (2.8) then becomes a random variable which needs to be updated using a Bayesian filter. The theoretical background, presented in Chapter 2, directly extends to this setup and leads to the highly interesting observation of "dual control" in MPC. Unfortunately, as already remarked in (Feldbaum, 1961a), the finite horizon optimal control problem to be solved is in general highly intractable. While various researchers have worked on deriving computationally tractable approximations, so far the problem could not be simplified such that the main properties were retained and rigorous guarantees on the solution that enable its application in MPC could be given.

It is hence only natural to simplify the setup and consider first an unknown but bounded, set-membership assumption on the uncertainty, removing the additional information of the distribution. In the following, we introduce a mathematically rigorous and computationally tractable framework for online system identification in model predictive control. Analogous to a Bayesian update rule of the conditional distribution, we recursively update a point estimate as well as bounding set. The resulting MPC algorithm represents a solution for constrained control of uncertain systems, where the conservatism is reduced through online learning.

The remainder of this chapter is structured as follows. In Section 4.1, we introduce the problem setup. Thereafter, in Section 4.2 the MPC framework is presented including the online system identification and the state prediction. The closed-loop properties are summarized in Section 4.3 and the chapter concludes with a numerical example followed by a discussion of the results.

This chapter is based on (Lorenzen et al., 2017a, 2018).

4.1 Problem setup

Similar to the previous chapter, consider a linear, discrete-time system

$$x_{k+1} = A(\theta)x_k + B(\theta)u_k + w_k \tag{4.1}$$

with state $x_k \in \mathbb{R}^n$, input $u_k \in \mathbb{R}^m$ and additive disturbance $w_k \in \mathbb{W} \subset \mathbb{R}^n$. Unlike the previous chapter, the system matrices $A : \mathbb{R}^{n_p} \to \mathbb{R}^{n\times n}$ and $B : \mathbb{R}^{n_p} \to \mathbb{R}^{n\times m}$ are affine functions in an unknown but *constant* parameter $\theta = \theta^* \in \mathbb{R}^{n_p}$. We impose the following assumption on the disturbance and uncertainty.

Assumption 4.1 (Disturbance and Uncertainty). *The disturbance set $\mathbb{W}$ is a bounded, convex polytope given by*

$$\mathbb{W} = \{w \in \mathbb{R}^n \mid H_w w \leq h_w\}$$

with $H_w \in \mathbb{R}^{q_w\times n}$ and $h_w \in \mathbb{R}^{q_w}$. The system matrices depend affinely on the parameter vector $\theta \in \mathbb{R}^{n_p}$

$$(A(\theta), B(\theta)) = (A_0, B_0) + \sum_{i=1}^{n_p}(A_i, B_i)[\theta]_i \tag{4.2}$$

and a bounded parameter set given by

$$\Theta = \{\theta \in \mathbb{R}^p \mid H_\theta\theta \leq h_\theta\}, \tag{4.3}$$

which contains the true parameter vector θ^* is known.

Remark 4.1. As shown in (Lorenzen et al., 2018), the following framework directly extends to slowly changing parameters. Yet, for clarity of presentation, we restrict ourselves to constant parameters.

Since no distributional information is assumed on the disturbance and uncertainty, similar to Section 2.1, hard constraints on the state and input $(x_k, u_k) \in \mathbb{Z}$ are imposed on the system. With given matrices $F \in \mathbb{R}^{q_c \times n}$ and $G \in \mathbb{R}^{q_c \times m}$, the constraint set $\mathbb{Z}$ is assumed to be given by a bounded, convex polytope

$$\mathbb{Z} = \{(x, u) \in \mathbb{R}^n \times \mathbb{R}^m \mid Fx + Gu \leq 1\}. \tag{4.4}$$

As before, the performance of the control system is measured by the running cost $\ell(x, u) = \|x\|_Q^2 + \|u\|_R^2$ with $Q \in \mathbb{R}^{n \times n}$, $Q \succ 0$ and $R \in \mathbb{R}^{m \times m}$, $R \succ 0$.

In the following, an adaptive MPC algorithm is derived, which achieves the objective of finding a stabilizing control law for system (4.1) such that the constraints $(x_k, u_k) \in \mathbb{Z}$ are satisfied robustly and the closed-loop cost is minimized by consistently updating the model uncertainty based on the state and input history. The following paragraph summarizes the main idea by suitably extending the basic algorithm presented in Chapter 2.

Indirect adaptive MPC

To solve the control task, we formulate a tube MPC algorithm with recursive online parameter estimation. As in the previous chapter, the requirements for constraint satisfaction and stability are considered separately. Specifically, the former requires a bounding set Θ_k of the unknown parameter θ^*. In contrast to known tube MPC algorithms, for stability results, in particular to derive a closed-loop gain from the disturbance to the state, a point estimate $\hat{\theta}_k$ is employed in a certainty equivalence approach.

Based on the disturbance set $\mathbb{W}$ and updated parameter membership set Θ_k, predicted states $x_{l|k}$ are bounded by sets $\mathbb{X}_{l|k}$ satisfying

$$\mathbb{X}_{0|k} \ni x_k, \tag{4.5a}$$

$$\mathbb{X}_{l+1|k} \ni A(\theta)x + B(\theta)u_{l|k}(x) + w \qquad \forall x \in \mathbb{X}_{l|k}, w \in \mathbb{W}, \theta \in \Theta_k. \tag{4.5b}$$

The sequence $(\mathbb{X}_{l|k})_{l \in \mathbb{N}_0^N}$ is called an admissible state tube if additionally

$$x \times u_{l|k}(x) \in \mathbb{Z} \qquad \forall x \in \mathbb{X}_{l|k}. \tag{4.6}$$

With a suitably chosen input parameterization, cost function J_N, and terminal constraint set $\mathbb{X}_f$, the MPC finite horizon optimal control problem to be solved at each sampling time k is given by

$$V_N(x_k, \hat{\theta}_k, \Theta_k) = \min_{\mathbf{u}_{N|k}} J_N(x_k, \hat{\theta}_k, \mathbf{u}_{N|k})$$
$$\text{s.t. } (4.5),\ (4.6),\ \mathbb{X}_{N|k} \subseteq \mathbb{X}_f.$$

Similar to the previously presented MPC algorithms, given x_k and a minimizer $\mathbf{u}^*_{N|k}$, the adaptive MPC control law is defined by $\kappa(x_k) = u^*_{0|k}$. Additionally, at each iteration the parameter estimates $\hat{\theta}_k$ and Θ_k are updated recursively based on the previously applied input u_{k-1} and state measurement x_k.

In the following, suitable choices for the system identification, the cost function and the parameterization of the state tube will be derived such that a computationally tractable optimization program is obtained and the desired closed-loop properties can be proven.

4.2 Online identification and state prediction

In this section, first an appropriate choice for the system identification is presented, specifically a set-membership update as well as point estimate. Based thereon, a suitable choice for the tube parameterization is made, which allows to derive a computationally tractable reformulation of the optimization program. In this way, a basic framework to rigorously guarantee constraint satisfaction and recursive feasibility in an adaptive receding horizon control algorithm is obtained. Finally, we introduce a certainty equivalence cost based on the point estimate and contrast this choice with existing results.

Parameter estimation

While many parameter estimation schemes have been presented in the literature, most are not suitable for a rigorous adaptive MPC design. In particular, to prove constraint satisfaction, recursive feasibility, and closed-loop stability, a non-increasing outer bound for the parameter membership set as well as a finite gain from the disturbance to the prediction error are necessary. For the former, we introduce a set-membership parameter estimation for the state space description (4.1), (4.2). In order to non-conservatively guarantee constraint satisfaction, this yields a tight polytopic parameter bounding set at each time step. For the latter, a projected least mean squares algorithm to update the point estimates of the parameter vector is derived, which, with a suitable cost function, leads to a finite closed-loop gain from the additive disturbance to the system state x_k.

For the system identification, we rewrite the state-space model and define the regressor matrix $D(x, u) \in \mathbb{R}^{n \times p}$ for a given state x and input u by

$$D(x, u) = \left[A_1 x + B_1 u,\ A_2 x + B_2 u,\ \dots\ ,\ A_{n_p} x + B_{n_p} u\right].$$

Furthermore, the shorthand notation $D_k = D(x_k, u_k)$, and $d_k = A_0 x_{k-1} + B_0 u_{k-1} -$

x_k will be used. Note that D_k and d_k depend affinely on the state and input variables.

Exact membership set At time k, given x_k as well as the previous state x_{k-1} and input u_{k-1}, the set of admissible parameters satisfying the system dynamics (4.1) is given by the polytope

$$\begin{aligned} \Delta_k &= \{\theta \in \mathbb{R}^p \mid x_k - (A(\theta)x_{k-1} + B(\theta)u_{k-1}) \in \mathbb{W}\} \\ &= \{\theta \in \mathbb{R}^p \mid -H_w D_{k-1}\theta \leq h_w + H_w d_k\}. \end{aligned} \tag{4.7}$$

With this, the membership set Θ_k of the uncertain parameter θ can be updated recursively. Defining $\Theta_0 = \Theta$, an exact update is given by

$$\Theta_k = \Theta_{k-1} \cap \Delta_k. \tag{4.8}$$

The following lemma summarizes the relevant properties of the membership set.

Lemma 4.2 (Membership set). *The set Θ_k is a convex, polytopic set explicitly given in half-space form*

$$\Theta_k = \{\theta \in \mathbb{R}^p \mid H_{\theta_k}\theta \leq h_{\theta_k}\} \tag{4.9}$$

with $H_{\theta_k} \in \mathbb{R}^{q_k \times n_p}$ and $h_{\theta_k} \in \mathbb{R}^{q_k}$. Furthermore, $\theta^* \in \Theta_k$ and $\Theta_{k+1} \subseteq \Theta_k$.

While the intersection of polytopes given in half-space representation is easy to implement, redundant constraints should be removed in each iteration. Although this is not necessary from a theoretical perspective, it decreases the computational load of the MPC algorithm. Note that redundant constraints in (4.9) can be removed efficiently by solving a series of linear programs (Blanchini and Miani, 2015, Section 3.3).

Bounded complexity update Under the update law (4.8), no upper bound on the number of non-redundant half-spaces defining Θ_k in (4.9) can be given. In fact, it is well known that the complexity of the set obtained through an exact set-membership estimate can grow without bound, making it unsuitable for an MPC algorithm. In the following, based on the previous paragraph, an update rule which provides a fixed complexity and under which Lemma 4.2 remains valid will be proposed.

In the literature, several methods for bounded complexity set updates by means of simpler regions have been presented, including using ellipsoids (Fogel and Huang, 1982), parallelotopes (Chisci et al., 1998) or more general polytopes (Veres et al., 1999). Yet, most do not guarantee the necessary property of the sets being non-increasing, i.e., $\Theta_{k+1} \subseteq \Theta_k$. All algorithms employing polytopic sets, as discussed here, limit the complexity by explicitly bounding the number of half-spaces. In contrast, the following set estimate is based on having a polytopic outer bound with a priori chosen orientation of the half spaces. Similar to algorithms using ellipsoids, this implies a constant complexity, however with an increased flexibility.

Let the parameter membership set be given by

$$\Theta_k = \{\theta \in \mathbb{R}^p \mid H_\theta \theta \leq h_{\theta_k}\}, \tag{4.10}$$

where $H_\theta \in \mathbb{R}^{q \times n_p}$ is an a priori chosen matrix whose rows define the q normal directions of the facets of the polytope Θ_k. The right hand side h_{θ_k} is updated at each iteration with $h_{\theta_0} \in \mathbb{R}^q$ satisfying $\Theta \subseteq \Theta_0$ and $h_{\theta_{k+1}}$ given by the minimizer of the linear optimization program

$$\begin{aligned} \min_{h \in \mathbb{R}^q, \Lambda \in \mathbb{R}^{q \times p_w}_{\geq 0}} \quad & \mathbb{1}^\top h \\ \text{s.t.} \quad & \Lambda \begin{bmatrix} H_\theta \\ -H_w D_k \end{bmatrix} = H_\theta \\ & \Lambda \begin{bmatrix} h_{\theta_k} \\ h_w + H_w d_{k+1} \end{bmatrix} \leq h \end{aligned} \tag{4.11}$$

where $p_w = q + q_w$. The objective of optimization (4.11) is to maximize the area while satisfying the required subset equations.

Similar to Lemma 4.2, the main properties of the proposed set update can be summarized as follows. Loosely speaking, the estimated membership set is non-increasing, while always containing the true parameter.

Lemma 4.3. *Let Θ_k be given by (4.10) with $h_{\theta_0} \in \mathbb{R}^q$ such that $\Theta \subseteq \Theta_0$ and h_{θ_k} for $k > 0$ being a minimizer of (4.11). Then $\theta^* \in \Theta_k$ and*

$$\Theta_k \cap \Delta_{k+1} \subseteq \Theta_{k+1} \subseteq \Theta_k. \tag{4.12}$$

Proof. The constraints in (4.11) are equivalent to $\Theta_k \cap \Delta_{k+1} \subseteq \Theta_{k+1}$ (Kouvaritakis and Cannon, 2016, Lemma 5.6), which by induction implies $\theta^* \in \Theta_k$ for all $k \geq 0$. Furthermore $\Lambda = [I \ 0]$, $h = h_{\theta_k}$ is a feasible solution for (4.11) and since the constraints on the individual rows of $[h, \Lambda]$ are decoupled, it necessarily holds that $h^* \leq h_{\theta_k}$, which implies $\Theta_{k+1} \subseteq \Theta_k$. ∎

Remark 4.4. To achieve a tighter set estimate, instead of an update in each iteration based on only Δ_k, the update (4.11) could be performed based on the past n_u

iterations using $\Delta_k, \ldots, \Delta_{k-n_u}$ in a moving window or updating Θ_k only every n_u iterations, as proposed in (Chisci et al., 1998).

To decrease the computational load, similar to (Veres et al., 1999), the update of h_{θ_k} can be restricted to those entries $[h_{\theta_k}]_i$ of facets i that are closest to facets in Δ_k in an inner product sense. Yet, we emphasize that this update heuristic does not always yield the best reduction in size as not necessarily the facet that leads to the largest reduction is updated.

Point estimate In most publications on linear adaptive MPC, a Kalman filter or Recursive Least Squares is employed for parameter estimation. In contrast, we propose an adapted Least Mean Squares (LMS) filter in order to rigorously derive closed-loop stability guarantees. While the LMS filter shows a slower convergence, a finite gain from the disturbance to the prediction error can be proven. In fact, for FIR models, in (Hassibi et al., 1993), the LMS filter has been shown to be an $\mathcal{H}_\infty$ optimal map from the disturbance to the prediction error.

Given a parameter estimate $\hat{\theta}_k$, state x_k and input u_k, denote the predicted state by $\hat{x}_{1|k} = A(\hat{\theta}_k)x_k + B(\hat{\theta}_k)u_k$ and the prediction error by

$$\tilde{x}_{1|k} = A(\theta^*)x_k + B(\theta^*)u_k - \hat{x}_{1|k}. \tag{4.13}$$

With given $\hat{\theta}_0 \in \Theta$ and parameter update gain $\mu \in \mathbb{R}_{>0}$ chosen such that $\frac{1}{\mu} > \sup_{(x,u)\in\mathbb{Z}} \|D(x,u)\|^2$, the parameter estimate $\hat{\theta}_k$ is defined recursively by

$$\begin{aligned} \hat{\theta}_k^- &= \hat{\theta}_{k-1} + \mu D(x_{k-1}, u_{k-1})^\top \left(x_k - \hat{x}_{1|k-1}\right) \\ \hat{\theta}_k &= \Pi_\Theta\left(\hat{\theta}_k^-\right), \end{aligned} \tag{4.14}$$

where $\Pi_\Theta(\theta_0) := \arg\min_{\theta\in\Theta} \|\theta - \theta_0\|$ denotes the Euclidean projection of a point $\theta_0 \in \mathbb{R}^p$ onto the set Θ.

The following lemma summarizes the relevant properties, which will be used in the following stability analysis of the MPC closed loop. In particular the parameter estimate is contained in the a priori given parameter set and the gain from the disturbance to the prediction error is bounded by 1.

Lemma 4.5 (Point estimate). *If* $\sup_{k\in\mathbb{N}} \|x_k\| < \infty$, $\sup_{k\in\mathbb{N}} \|u_k\| < \infty$, *the parameter estimate* $\hat{\theta}_k$ *is bounded, consistent with the prior parameter set, i.e.,* $\hat{\theta}_k \in \Theta$, *and*

$$\sup_{m\in\mathbb{N}, w_k\in\mathbb{W}, \hat{\theta}_0\in\Theta} \frac{\sum_{k=0}^m \|\tilde{x}_{1|k}\|^2}{\frac{1}{\mu}\|\hat{\theta}_0 - \theta^*\|^2 + \sum_{k=0}^m \|w_k\|^2} \leq 1.$$

Proof. Boundedness of $\hat{\theta}_k$ and $\hat{\theta}_k \in \Theta$ follow trivially from the set update (4.7), (4.8) and projection in (4.14). To prove the bound on the prediction error consider

$$\begin{aligned}
&\frac{1}{\mu}\|\hat{\theta}_{k+1} - \theta^*\|^2 - \frac{1}{\mu}\|\hat{\theta}_k - \theta^*\|^2 \\
&\leq \frac{1}{\mu}\|\hat{\theta}^-_{k+1} - \theta^*\|^2 - \frac{1}{\mu}\|\hat{\theta}_k - \theta^*\|^2 \\
&= \frac{1}{\mu}\|\hat{\theta}^-_{k+1} - \hat{\theta}_k\|^2 + \frac{2}{\mu}(\hat{\theta}^-_{k+1} - \hat{\theta}_k)^\top(\hat{\theta}_k - \theta^*) \\
&= \frac{1}{\mu}\|\mu D_k^\top(\tilde{x}_{1|k} + w_k)\|^2 + 2(\tilde{x}_{1|k} + w_k)^\top D_k(\hat{\theta}_k - \theta^*) \\
&\leq (\mu\|D_k\|^2 - 1)\|\tilde{x}_{1|k} + w_k\|^2 - \|\tilde{x}_{1|k}\|^2 + \|w_k\|^2 \\
&\leq -\|\tilde{x}_{1|k}\|^2 + \|w_k\|^2
\end{aligned} \tag{4.15}$$

with the above introduced abbreviation $D_k = D(x_k, u_k)$. The first inequality follows from non-expansiveness of the projection operator and $\theta^* \in \Theta$. In the second equality and inequality we use (4.14) with $x_{k+1} - \hat{x}_{1|k} = \tilde{x}_{1|k} + w_k$, $\tilde{x}_{1|k} = D_k(\theta^* - \hat{\theta}_k)$, and completion of squares. Summing (4.15) from $k = 0$ to m yields

$$\frac{1}{\mu}\|\hat{\theta}_{m+1} - \theta^*\|^2 + \sum_{k=0}^{m}\|\tilde{x}_{1|k}\|^2 \leq \sum_{k=0}^{m}\|w_k\|^2 + \frac{1}{\mu}\|\hat{\theta}_0 - \theta^*\|^2$$

which proves the claim. ∎

An important corollary of Lemma 4.5 is that the prediction error converges to zero if the disturbance is a square-summable sequence.

Corollary 4.6. *If* $\sup_{k\in\mathbb{N}}\|x_k\| < \infty$, $\sup_{k\in\mathbb{N}}\|u_k\| < \infty$, *and* $\sum_{k=0}^{\infty}\|w_k\|^2 < \infty$, *the prediction error converges asymptotically to the origin, i.e.,* $\lim_{k\to\infty}\|\tilde{x}_{1|k}\| = 0$.

Remark 4.7. In order to increase the convergence speed, in (4.14) a projection onto the set Θ_k could be used without changing the result of Lemma 4.5. Projecting the parameter estimate onto a trust region is common practice for robust adaptive control algorithms (Åström and Wittenmark, 2008). Furthermore, note that the update gain μ is non-increasing with an increasing size of $\mathbb{Z}$ which is in line with common recommendations on stabilizing indirect adaptive control, as discussed, e.g., in (Bitmead et al., 1990, Chapter 7.4).

State prediction and constraint reformulation

As before, to obtain a finite dimensional optimization program, we introduce an input parameterization[1]

$$u_{l|k}(x) = Kx + v_{l|k} \tag{4.16}$$

with decision variables $\mathbf{v}_{N|k} = (v_{l|k})_{l \in \mathbb{N}_0^{N-1}}$, $v_{l|k} \in \mathbb{R}^m$ and constant feedback gain $K \in \mathbb{R}^{m \times n}$. The following assumption is made on the prestabilizing gain K in order to cope with state predictions under parametric uncertainty.

Assumption 4.2 (Prestabilization). *The feedback gain K is chosen such that $A_{cl}(\theta) = A(\theta) + B(\theta)K$ is stable for all $\theta \in \Theta$.*

While this assumption, compared to most adaptive control literature, is conservative, it is necessary in the presence of hard state constraints. Note that, given the prior parameter set Θ, the gain K can be determined by standard robust control methods.

Additionally, to obtain a computationally tractable optimization program, the state tube cross sections $\mathbb{X}_{l|k}$ are restricted to translations and dilations of a fixed polytope $\mathbb{X}_0$. In the context of robust MPC, this parameterization has previously been introduced in (Langson et al., 2004) as well as (Raković et al., 2012).

In the following, let $\mathbb{X}_0$ be given by $\mathbb{X}_0 = \{x \in \mathbb{R}^n \mid H_x x \leq \mathbb{1}\}$, $H_x \in R^{u \times n}$ with vertices $\{x^1, \ldots, x^v\}$ and define the bounding sets

$$\begin{aligned} \mathbb{X}_{l|k} &= \{z_{l|k}\} \oplus \alpha_{l|k}\mathbb{X}_0 \\ &= \{x \in \mathbb{R}^n \mid H_x(x - z_{l|k}) \leq \alpha_{l|k}\mathbb{1}\} \\ &= \{z_{l|k}\} \oplus \alpha_{l|k}\,\mathrm{co}\{x^1, x^2, \ldots, x^v\}. \end{aligned} \tag{4.17}$$

The MPC decision variables defining the state tube are thus given by $\mathbf{z}_{N|k} = (z_{l|k})_{l \in \mathbb{N}_0^N}$, $z_{l|k} \in \mathbb{R}^n$, and $\boldsymbol{\alpha}_{N|k} = (\alpha_{l|k})_{l \in \mathbb{N}_0^N}$, $\alpha_{l|k} \in \mathbb{R}_{\geq 0}$.

This explicit description of the sets $\mathbb{X}_{l|k}$ in both vertex *and* half-space form can be exploited to reformulate the set prediction (4.5) and the constraints (4.6) as linear constraints. In particular, this is done without further offline computations, but directly taking into account the online updated parameter set Θ_k. This result is formally stated in the following proposition, which is the cornerstone for a computationally tractable reformulation of the basic adaptive MPC optimization program.

[1] Similar to the previous section, for clarity of presentation we restrict ourselves to this input parameterization. To reduce the conservatism, a different prestabilizing input parameterization that is affine in the decision variables could be used without changing the following results.

For a concise presentation, with $x^j_{l|k} = z_{l|k} + \alpha_{l|k} x^j$, $u^j_{l|k} = u_{l|k}(x^j_{l|k})$ define $D^j_{l|k} = D(x^j_{l|k}, u^j_{l|k})$ and $d^j_{l|k} = A_0 x^j_{l|k} + B_0 u^j_{l|k} - z_{l+1|k}$. Furthermore, for all $i \in \mathbb{N}_1^u$ let $[\bar{w}]_i = \max_{w \in \mathbb{W}} [H_x]_i w$ and for all $i \in \mathbb{N}_1^c$ let $[\bar{f}]_i = \max_{x \in \mathbb{X}_0} [F + GK]_i x$ with F, G being the constraint matrices in (4.4).

Proposition 4.8 (Prediction tube). *Let the input $\mathbf{u}_{N|k}$ be parameterized as in (4.16) and $\{\mathbb{X}_{l|k}\}_{l \in \mathbb{N}_0^N}$ as in (4.17) with decision variables $\mathbf{z}_{N|k}$, $\boldsymbol{\alpha}_{N|k}$, and $\mathbf{v}_{N|k}$.*

Equations (4.5), (4.6) are satisfied if and only if for all $j \in \mathbb{N}_1^v$, $l \in \mathbb{N}_0^{N-1}$ there exists $\Lambda^j_{l|k} \in \mathbb{R}_{\geq 0}^{u \times q_k}$ satisfying

$$(F + GK) z_{l|k} + G v_{l|k} + \alpha_{l|k} \bar{f} \leq \mathbb{1} \tag{4.18a}$$

$$- H_x z_{0|k} - \alpha_{0|k} \mathbb{1} \leq -H_x x_k \tag{4.18b}$$

$$\Lambda^j_{l|k} h_{\theta_k} + H_x d^j_{l|k} - \alpha_{l+1|k} \mathbb{1} \leq -\bar{w} \tag{4.18c}$$

$$H_x D^j_{l|k} = \Lambda^j_{l|k} H_{\theta_k}. \tag{4.18d}$$

Proof. Inequality (4.6) is equivalent to

$$(F + GK) z_{l|k} + G v_{l|k} + \alpha_{l|k} (F + GK) x \leq \mathbb{1} \quad \forall x \in \mathbb{X}_0$$

which is equivalent to (4.18a) when maximizing over $x \in \mathbb{X}_0$.

Inequality (4.5a) is equivalent to (4.18b), and (4.5b) is equivalent to (4.18c), (4.18d) as shown by the following reformulation.

$$
\begin{aligned}
& \mathbb{X}_{l+1|k} \supseteq A_{cl}(\theta) \mathbb{X}_{l|k} \oplus B(\theta) v_{l|k} \oplus \mathbb{W} && \forall \theta \in \Theta_k \\
\Leftrightarrow\ & H_x (A_{cl}(\theta) x + B(\theta) v_{l|k} + w - z_{l+1|k}) \leq \alpha_{l+1|k} \mathbb{1} && \forall \theta \in \Theta_k, w \in \mathbb{W}, \\
& && x \in \mathbb{X}_{l|k} \\
\Leftrightarrow\ & H_x (A_{cl}(\theta)(z_{l|k} + \alpha_{l|k} x^j) + B(\theta) v_{l|k} - z_{l+1|k}) - \alpha_{l+1|k} \mathbb{1} \leq -\bar{w} && \forall \theta \in \Theta_k, j \in \mathbb{N}_1^v \\
\Leftrightarrow\ & \max_{\theta \in \Theta_k} \left\{ H_x (A_{cl}(\theta)(z_{l|k} + \alpha_{l|k} x^j) + B(\theta) v_{l|k}) \right\} && \forall j \in \mathbb{N}_1^v \\
& \qquad - H_x z_{l+1|k} - \alpha_{l+1|k} \mathbb{1} \leq -\bar{w} && \\
\Leftrightarrow\ & \max_{\theta \in \Theta_k} \left\{ H_x D^j_{l|k} \theta \right\} + H_x d^j_{l|k} - \alpha_{l+1|k} \mathbb{1} \leq -\bar{w} && \forall j \in \mathbb{N}_1^v \\
\Leftrightarrow\ & \left\{ \begin{array}{l} \Lambda^j_{l|k} h_{\theta_k} + H_x d^j_{l|k} - \alpha_{l+1|k} \mathbb{1} \leq -\bar{w} \\ H_x D^j_{l|k} = \Lambda^j_{l|k} H_{\theta_k} \\ \Lambda^j_{l|k} \in \mathbb{R}_{\geq 0}^{u \times q_k} \end{array} \right\} && \forall j \in \mathbb{N}_1^v
\end{aligned}
$$

The first equivalence follows from $x \in \mathbb{X}_{l+1|k}$ being equivalent to $H_x(x - z_{l+1|k}) \leq \alpha_{l+1|k}\mathbb{1}$. The second one follows from the left hand side being convex in x for given θ so that the inequality holds for all $x \in \mathbb{X}_{l|k}$ if and only if it holds for the vertices of $\mathbb{X}_{l|k}$. In the third equivalence the maximisation is to be understood row-wise and the last equivalence follows from strong duality in linear programming. Specifically, considering row $i \in \mathbb{N}_1^u$ in the maximisation, which is finite since Θ_k is compact, we have

$$\begin{aligned}
&\max_{\theta \in \Theta_k} \left\{ [H_x]_i D^j_{l|k}\theta \right\} \\
&= \min_{\lambda_i \in \mathbb{R}^{q_k}_{\geq 0}} \max_{\theta} \left\{ [H_x]_i D^j_{l|k}\theta + \lambda_i^\top (h_{\theta_k} - H_{\theta_k}\theta) \right\} \\
&= \min_{\lambda_i \in \mathbb{R}^{q_k}_{\geq 0}} \lambda_i^\top h_{\theta_k} \\
&\qquad \text{s.t. } [H_x]_i D^j_{l|k} - \lambda_i^\top H_{\theta_k} = 0.
\end{aligned}$$

Since the minimization in the inequality constraint can be removed, this concludes the proof. ∎

Remark 4.9. The linear equality constraint (4.18d) can equivalently be written in the usual vector notation as

$$\begin{bmatrix} H_x A_1 & H_x A_1 x^j & H_x B_1 & -I_u \otimes [H_\theta^\top]_1 \\ H_x A_2 & H_x A_2 x^j & H_x B_2 & -I_u \otimes [H_\theta^\top]_2 \\ & & \cdots & \\ H_x A_p & H_x A_p x^j & H_x B_p & -I_u \otimes [H_\theta^\top]_p \end{bmatrix} \begin{bmatrix} z_{l|k} \\ \alpha_{l|k} \\ u_{l|k} \\ \mathrm{vec}((\Lambda^j_{l|k})^\top) \end{bmatrix} = 0.$$

Lemma 4.2 and Proposition 4.8 provide a computationally tractable, basic framework for a predictive control algorithm with model adaption and robust constraint satisfaction. In the following, this is completed with a suitable terminal constraint as well as a convex, positive definite objective function to derive a stabilizing, adaptive MPC algorithm.

Terminal constraint and objective function

To derive a stabilizing adaptive MPC algorithm, we draw upon the basic results presented in Chapter 2. In particular, to prove recursive feasibility, we assume existence of a suitable terminal set, which is robust forward invariant for the state tube dynamics. To prove stability, we employ a terminal cost which satisfies a variant of the basic stability assumption.

Terminal set The following assumption on the terminal set $\mathbb{X}_f$ is analogous to assumptions on a terminal set in robust MPC with homothetic tubes. Similar to the remark on Assumption 4.2, this is conservative compared to most results in adaptive control, yet necessary if state constraints should be satisfied robustly. If only input constraints are present, the terminal constraint can be omitted by setting the prestabilizing feedback gain K to zero.

Assumption 4.3 (Terminal set). *There exists a nonempty terminal set $\mathbb{X}_f = \{(z, \alpha) \in \mathbb{R}^n \times \mathbb{R}_{\geq 0} \mid H_T z + h_T \alpha \leq \mathbb{1}\}$ such that $(x, Kx) \in \mathbb{Z}$ for all $x \in \{z\} \oplus \alpha\mathbb{X}_0$, $(z, \alpha) \in \mathbb{X}_f$ and for all $\theta \in \Theta$ we have*

$$(z, \alpha) \in \mathbb{X}_f \implies \exists (z^+, \alpha^+) \in \mathbb{X}_f \text{ s.t. } A_{cl}(\theta)(\{z\} \oplus \alpha\mathbb{X}_0) \oplus \mathbb{W} \subseteq \{z^+\} \oplus \alpha^+\mathbb{X}_0.$$

Remark 4.10. Analogous to the remark in Chapter 3, through explicitly considering the dynamics of (z, α), the terminal set $\mathbb{X}_f$ can be computed recursively by standard algorithms to determine a robust forward invariant set, cf. (Raković and Cheng, 2013). Yet, note that standard algorithms, as presented in, e.g., (Blanchini and Miani, 2015), involve projection of polytopes, which can be demanding. Here, this can be significantly alleviated through an additional constraint $z = 0$ in $\mathbb{X}_f$ and determining only a suitable α satisfying Assumption 4.3.

Objective function Given the point estimate $\hat{\theta}_k$, we can define a certainty equivalence cost. With the result in Lemma 4.5, this leads to a finite ℓ_2 gain from the disturbance to the state of the closed-loop system. To satisfy the basic stability assumption with running cost $\ell(x, u) = \|x\|_Q^2 + \|u\|_R^2$ and local control law $\kappa_f(x) = Kx$, the terminal cost is chosen as $V_f(x) = \|x\|_P^2$ with P satisfying

$$A_{cl}(\theta)^\top P A_{cl}(\theta) + Q + K^\top R K \preceq P \quad \forall \theta \in \Theta. \tag{4.19}$$

Summarizing, the finite horizon MPC cost function is then given by

$$J_N(x_k, \hat{\theta}_k, \mathbf{v}_{N|k}) = \sum_{l=0}^{N-1} \|\hat{x}_{l|k}\|_Q^2 + \|u_{l|k}(\hat{x}_{l|k})\|_R^2 + \|\hat{x}_{N|k}\|_P^2 \tag{4.20}$$

where $\hat{x}_{l|k}$, $\hat{u}_{l|k}$ are implicitly defined by

$$\begin{aligned} \hat{x}_{l+1|k} &= A(\hat{\theta}_k)\hat{x}_{l|k} + B(\hat{\theta}_k)u_{l|k}(\hat{x}_{l|k}), \qquad x_{0|k} = x_k, \\ u_{l|k}(\hat{x}_{l|k}) &= K\hat{x}_{l|k} + v_{l|k}. \end{aligned} \tag{4.21}$$

We emphasize that, unlike in previous publications, e.g., (Marafioti et al., 2014) or (Di Cairano, 2016), the parameter estimate is directly used in the predictions (4.21). To prove stability, often the parameter is changed only after a certain dwell time or alternatively changed only at the end of the prediction horizon, leading to the prediction equation $\hat{x}_{l+1|k} = A(\hat{\theta}_{k-N+l})\hat{x}_{l|k} + B(\hat{\theta}_{k-N+l})u_{l|k}(\hat{x}_{l|k})$. While both allows for a simpler analysis, it slows down the convergence of the prediction error.

Remark 4.11. For a stabilizing MPC algorithm, the certainty equivalence cost (4.20) could be replaced by a min-max cost

$$\ell(z_{l|k}, \alpha_{l|k}, v_{l|k}) = \max_{x \in \mathbb{X}_{l|k}} \left\{ \|Qx\|_\infty + \|Ru_{l|k}(x)\|_\infty \right\}. \tag{4.22}$$

The advantage of this alternative formulation is that the cost depends only on the set estimate Θ_k and the point estimate $\hat{\theta}_k$ for the parameter can be omitted. With a suitable terminal cost, and under an additional, technical assumption, practical stability of the closed loop under this min-max cost has been proven in (Lorenzen et al., 2017a). However, the disadvantage of a min-max approach is that the ultimate bound on the state depends on the set $\mathbb{W}$ whereas the approach proposed in this paper, i.e., a certainty equivalence cost, allows the ℓ_2 norm of the state to be bounded in terms of the ℓ_2 norm of the actually realized disturbance sequence.

4.3 Adaptive MPC algorithm and closed-loop properties

In summary, a suitable choice of the system identification as well as parameterization of the state tube has led to a computationally tractable reformulation of the MPC optimization in terms of a linearly constrained quadratic program. In the following, we summarize the complete algorithm and thereafter analyze its control theoretic properties. Apart from proving closed-loop stability and constraint satisfaction, we discuss persistence of excitation and convergence of the parameter estimates.

In order to simplify notation, we denote the decision variables in the online optimization program by $\mathbf{d}_{N|k} = (\mathbf{z}_{N|k}, \boldsymbol{\alpha}_{N|k}, \mathbf{v}_{N|k}, \boldsymbol{\Lambda}_{N|k})$ with $\boldsymbol{\Lambda}_{N|k} = (\Lambda^j_{l|k})_{j \in \mathbb{N}_1^v, l \in \mathbb{N}_0^{N-1}}$. For a given state x_k and parameter set Θ_k, the set of admissible decision variables of the adaptive MPC optimization is given by

$$\mathbb{D}(x_k, \Theta_k) = \{\mathbf{d}_{N|k} \mid (4.18), (z_{N|k}, \alpha_{N|k}) \in \mathbb{X}_f\}$$

and the set of feasible initial conditions is given by $\mathbb{X}_N(\Theta_k) = \{x \in \mathbb{R}^n \mid \mathbb{D}(x, \Theta_k) \neq \emptyset\}$. In contrast to the previous chapter, the set $\mathbb{X}_N$ depends on the estimated parameter set Θ_k highlighting its dependence on the online parameter estimation.

With this we can define the adaptive MPC algorithm as follows.

Algorithm 4.12 (Adaptive MPC for linear systems).
Offline: Choose a polytope $\mathbb{X}_0$, which determines the state tube parameterization. Compute a robustly stabilizing feedback gain K and terminal set $\mathbb{X}_f$ according to Assumption 4.2 and 4.3, respectively. Derive a terminal cost matrix P satisfying (4.19).
Online: For each time step $k = 0, 1, 2, \ldots$

1. Measure the current state x_k.
2. If $k > 0$ update the membership set Θ_k and point estimate $\hat{\theta}_k$ according to (4.8) or (4.11), and (4.14), respectively.
3. Determine the minimizer $\mathbf{d}^*_{N|k}$ of the linearly constrained quadratic program

$$\begin{aligned} \mathbf{d}^*_{N|k} = \arg\min_{\mathbf{d}_{N|k}} \; & J_N(x_k, \hat{\theta}_k, \mathbf{v}_{N|k}) \\ \text{s.t.} \; & \mathbf{d}_{N|k} \in \mathbb{D}(x_k, \Theta_k). \end{aligned} \tag{4.23}$$

4. Apply the adaptive MPC feedback value $\kappa(x_k) = Kx_k + v^*_{0|k}$.

Closed-loop properties

We can now state the following result, which shows that the closed-loop system resulting from application of the proposed adaptive MPC algorithm is finite gain ℓ_2 stable and the resulting trajectories satisfy the constraints robustly. Furthermore, building upon the results in Section 4.2, the estimated parameter set always contains the true parameter θ^*.

Theorem 4.13 (Adaptive MPC closed-loop properties). *Consider the adaptive MPC Algorithm 4.12 in closed loop with the linear system (4.1) and suppose Assumptions 4.1, 4.2, and 4.3 are satisfied. If $\theta^* \in \Theta$ and $x_0 \in \mathbb{X}_N(\Theta)$, then the MPC optimization (4.23) is feasible for all $k \in \mathbb{N}$, the true parameter is contained in the set estimate, i.e., $\theta^* \in \Theta_k$, and the point estimate satisfies the a priori bound $\hat{\theta}_k \in \Theta$. Furthermore, the state and input trajectories satisfy the hard constraints $x_k \times u_k \in \mathbb{Z}$ and the closed-loop system is finite gain ℓ_2 stable, i.e., there exist constants $c_0, c_1, c_2 \in \mathbb{R}_{>0}$ such that for all $K \in \mathbb{N}$*

$$\sum_{k=0}^{K} \|x_k\|^2 \leq c_0 \|x_0\|^2 + c_1 \|\hat{\theta}_0 - \theta^*\|^2 + c_2 \sum_{k=0}^{K} \|w_k\|^2.$$

Remark 4.14. Unlike nominal MPC, the adaptive MPC controller is not a static state feedback, but a dynamic controller with states $\hat{\theta}_k$ and Θ_k. The term $c_0\|x_0\|^2 + c_1\|\hat{\theta}_0 - \theta^*\|^2$ bounds the possible overshoot due to the initial condition of the plant *and* the controller.

Proof. We prove the first and second claim, feasibility and consistent parameter estimation, by induction.

By Assumption, $\theta^* \in \Theta_0$ and $x_0 \in \mathbb{X}_N(\Theta_0)$. Suppose that $\theta^* \in \Theta_{k_0}$ and $x_{k_0} \in \mathbb{X}_N(\Theta_{k_0})$ at time k_0.
Feasibility: For $l \in \mathbb{N}_0^{N-1}$, let $u_{l|k_0}(x) = Kx + v^*_{l|k_0}$ and $\mathbb{X}^*_{l+1|k_0}$ be a feasible input and admissible state tube trajectory satisfying the tube constraints (4.5), (4.6) and terminal constraint $\mathbb{X}_{N|k_0} \subseteq \mathbb{X}_f$[2]. For time k_0+1 and $l \in \mathbb{N}_0^{N-1}$ define the candidate input $\tilde{u}_{l|k_0+1}(x) = Kx + v^*_{l+1|k_0}$ with $v^*_{N|k_0} = 0$ and the candidate state tube $\tilde{\mathbb{X}}_{l|k_0+1} = \mathbb{X}^*_{l+1|k_0}$. Since $\tilde{\mathbb{X}}_{N-1|k_0+1} = \mathbb{X}^*_{N|k_0} \subseteq \mathbb{X}_f$ and $\Theta_{k_0} \subseteq \Theta$, by Assumption 4.3, there exists $\tilde{\mathbb{X}}_{N|k_0+1} \subseteq \mathbb{X}_f$ satisfying $A(\theta)x \oplus B(\theta)\tilde{u}_{N-1|k_0+1}(x) \oplus \mathbb{W} \subseteq \tilde{\mathbb{X}}_{N|k_0+1}$ for all $x \in \tilde{\mathbb{X}}_{N-1|k_0+1}$, $\theta \in \Theta_{k_0}$. By construction $(\tilde{u}_{l|k_0+1}, \tilde{\mathbb{X}}_{l|k_0+1})_{l\in\mathbb{N}_0^{N-1}}$ satisfy the constraints (4.5b), (4.6), and since $x_{k_0+1} = A(\theta^*)x_{k_0} + B(\theta^*)u_{k_0} + w_{k_0} \in \mathbb{X}^*_{1|k_0} = \tilde{\mathbb{X}}_{0|k_0+1}$, constraint (4.5a) is satisfied. By Proposition 4.8 this is equivalent to feasibility of (4.18) and hence $x_{k_0+1} \in \mathbb{X}_N(\Theta_{k_0})$ which implies $x_{k_0+1} \in \mathbb{X}_N(\Theta_{k_0+1})$ as $\Theta_{k_0+1} \subseteq \Theta_{k_0}$.

Parameter estimation: If x_{k_0} is bounded, the optimal solution $v^*_{0|k_0}$ and hence u_{k_0} and x_{k_0+1} are bounded (Rawlings et al., 2017). The claim then follows from Lemma 4.2 and 4.5.

The third claim, *constraint satisfaction*, is a direct corollary of $x_k \in \mathbb{X}_N(\Theta_k)$ for all k and Proposition 4.8.

To prove the *finite ℓ_2 gain* we take again $\tilde{v}_{l|k+1} = v^*_{l+1|k}$ for $l \in \mathbb{N}_0^{N-2}$ and $\tilde{v}_{N-1|k+1} = 0$ as a feasible, candidate input sequence at time $k+1$. Let $(\hat{x}_{l|k})_{l\in\mathbb{N}_1^N}$ and $(\hat{x}_{l-1|k+1})_{l\in\mathbb{N}_1^N}$ denote the corresponding predicted state trajectories, which evolve according to (4.21) with initial conditions x_k and x_{k+1}, respectively, and denote the difference by

$$\begin{aligned}\delta\hat{x}_{l|k} &= \hat{x}_{l-1|k+1} - \hat{x}_{l|k}\\ &= A_{cl}(\hat{\theta}_{k+1})^{l-1}(w_k + \tilde{x}_{1|k})\\ &\quad + \sum_{i=1}^{l-1} A_{cl}(\hat{\theta}_{k+1})^{l-1-i} D(\hat{x}_{i|k}, \hat{u}_{i|k})(\hat{\theta}_{k+1} - \hat{\theta}_k).\end{aligned}$$

[2]To simplify the notation in the proof, we write $\mathbb{X}_{N|k} \subseteq \mathbb{X}_f$ to denote $(z_{N|k}, \alpha_{N|k}) \in \mathbb{X}_f$.

By non-expansiveness of the projection operator, we have $\|\hat{\theta}_{k+1} - \hat{\theta}_k\| \leq \|\hat{\theta}^-_{k+1} - \hat{\theta}_k\|$, and using (4.14) together with $\frac{1}{\mu} > \sup_{(x,u)\in\mathbb{Z}} \|D(x,u)\|^2$ this leads to

$$\|\delta\hat{x}_{l|k}\| \leq \left(\sum_{i=0}^{l-1} \|A_{cl}(\theta_{k+1})^{l-i}\|\right) \|\tilde{x}_{1|k} + w_k\|.$$

The claim then follows by a standard argument. With the optimal value function $V_N(x_k, \hat{\theta}_k, \Theta_k) = J_N(x_k, \hat{\theta}_k, \mathbf{v}^*_{N|k})$ and $\bar{Q} = Q + K^\top RK$ consider

$$\begin{aligned}
&V_N(x_{k+1}, \hat{\theta}_{k+1}, \Theta_{k+1}) - V_N(x_k, \hat{\theta}_k, \Theta_k) \\
\leq& J_N(x_{k+1}, \hat{\theta}_{k+1}, \tilde{\mathbf{v}}_{N|k}) - V_N(x_k, \hat{\theta}_k, \Theta_k) \\
\leq& - \|x_k\|^2_Q - \|u_k\|^2_R + \sum_{l=0}^{N-2} \|\hat{x}_{l|k+1}\|^2_Q + \|\hat{u}_{l|k+1}\|^2_R \\
&+ \|\hat{x}_{N-1|k+1}\|^2_P - \left(\sum_{l=1}^{N-1} \|\hat{x}_{l|k}\|^2_Q + \|\hat{u}_{l|k}\|^2_R + \|\hat{x}_{N|k}\|^2_P\right) \\
\leq& - \|x_k\|^2_Q - \|u_k\|^2_R + \sum_{l=1}^{N-1} \varepsilon \left(\|\hat{x}_{l|k}\|^2_Q + \|\hat{u}_{l|k}\|^2_R\right) \\
&+ \varepsilon\|\hat{x}_{N|k}\|^2_P + \sum_{l=1}^{N-1} \left(1 + \frac{1}{\varepsilon}\right) \|\delta\hat{x}_{l|k}\|^2_{\bar{Q}} + \|\delta\hat{x}_{N|k}\|^2_P \\
\leq& - \|x_k\|^2_Q + \varepsilon V_N(x_k, \hat{\theta}_k, \Theta_k) + c_A\|\tilde{x}_{1|k} + w_k\|^2 \\
\leq& - c\|x_k\|^2 + c_A\|\tilde{x}_{1|k} + w_k\|^2.
\end{aligned} \tag{4.24}$$

The third inequality follows by Cauchy-Schwarz and Young's inequality which implies $\|\hat{x}_{l-1|k+1}\|^2_Q = \|\hat{x}_{l|k} + \delta\hat{x}_{l|k}\|^2_Q \leq (1+\varepsilon)\|\hat{x}_{l|k}\|^2_Q + (1+\frac{1}{\varepsilon})\|\delta x_{l|k}\|^2_Q$. Since for each $\theta \in \Theta$, $V_N(x, \theta, \Theta)$ is a continuous, piecewise quadratic function in x, cf. Bemporad et al. (2002), it can be upper bounded by a quadratic function on the feasible set. In particular for each $\theta \in \Theta$, there exists c_θ, such that $V_N(x, \theta, \Theta) = J_N(x, \theta, \mathbf{v}^*_N(x, \theta, \Theta)) < c_\theta\|x\|^2$, where $\mathbf{v}^*_N$ is the optimal solution of (4.23) with $x_k = x$, $\hat{\theta}_k = \theta$, $\Theta_k = \Theta$. Since $J_N(x, \cdot, \mathbf{v}^*_N(x, \theta, \Theta))$ is continuous, we can choose an $\varepsilon > 0$ such that $J_N(x, \tilde{\theta}, \mathbf{v}^*_N(x, \theta, \Theta)) \leq c_\theta\|x\|^2$ for each $\tilde{\theta} \in \mathbb{B}_\varepsilon(\theta)$ and each $\theta \in \Theta$. By compactness of Θ, there exists a finite collection $\{\theta^i\}_{i\in\mathcal{I}}$ with $\theta^i \in \Theta$ such that $\cup_{i\in\mathcal{I}}\mathbb{B}_\varepsilon(\theta^i) \supseteq \Theta$. Hence, $J_N(x_k, \hat{\theta}_k, \mathbf{v}^*_{N|k}) \leq \hat{c}_\theta\|x\|^2$ with $\hat{c}_\theta = \max_{i\in\mathcal{I}} c_{\theta^i}$. Thus there exist suitable constants $\varepsilon, c, c_A \in \mathbb{R}_{>0}$ such that the last two inequalities hold. The final result follows by summing over k, using Lemma 4.5, and again Young's inequality for $\|\tilde{x}_{1|k} + w_k\|^2$. ∎

Corollary 4.15 (Convergence of the closed loop). *Consider the adaptive MPC Algorithm 4.12 in closed loop with the linear system (4.1) and suppose Assumptions 4.1, 4.2, and 4.3 are satisfied. If $\sum_{k=0}^{\infty} \|w_k\|^2 < \infty$ and $x_0 \in \mathbb{X}_N(\Theta)$, then the solution of the closed-loop system converges asymptotically to the origin, i.e., $\lim_{k\to\infty} x_k = 0$.*

We have proved robust constraint satisfaction and a finite ℓ_2 gain for a computationally tractable MPC algorithm with online parameter adaption.

Parameter convergence and persistence of excitation

Unlike in a perfect dual control formulation, the MPC finite horizon optimal control program (4.23) does not take into account future parameter identification. Hence the optimal solution does not explicitly excite the system to better identify the parameters, but the learning is passive and rather "accidental". In classical adaptive control design, it is well known that this lack of excitation can lead to a drift in the parameter estimates, which might ultimately result in bursts or oscillatory behavior of the closed-loop system. For this reason, persistence of excitation (PE) has become one central property in the adaptive control literature, cf. (Narendra and Annaswamy, 2005). As an outer bound for the parameters is assumed to be known, for the presented adaptive MPC algorithm a PE condition was not necessary to prove stability. Yet, if it is satisfied, it allows to prove convergence of the parameter estimates, and thereby indirectly improves the closed-loop performance.

In the following, we recall persistence of excitation and derive results on parameter convergence under a PE assumption. Finally, we translate this assumption on the regressor matrix to a condition on the input vector, which can be directly included in the MPC optimization if the disturbance sequence $(w_k)_{k\in\mathbb{N}}$ is not assumed to be persistently exciting.

Throughout this section, we make the following linear independence assumption, which is necessary for uniquely estimating the parameters. If Assumption 4.4 is violated, one or more parameters can be equivalently expressed by a linear combination of the others. Note that it can be rewritten as a matrix rank condition and checked easily.

Assumption 4.4 (Linear independence). *The equation $\sum_{i=1}^{p} \lambda_i [A_i\ B_i] = 0$ has only the trivial solution $\lambda_i = 0$.*

Several definitions of persistence of excitation have been given in the literature. The following definition naturally generalizes the more standard definition of a regressor vector $\phi_k \in \mathbb{R}^p$ to a matrix valued regressor $\Phi_k \in \mathbb{R}^{p\times q}$. Note that for $\Phi = [\phi_{k,1}\ \dots\ \phi_{k,1}]$ the term $\Phi_k\Phi_k^\top$ is equivalent to $\sum_{i=1}^{q} \phi_{k,i}\phi_{k,i}^\top$.

Definition 4.16 (Persistence of excitation). *A regressor $(\Phi_k)_{k\in\mathbb{N}}$ with $\Phi_k \in \mathbb{R}^{p\times q}$ is persistently exciting if there exist positive constants α, β, P such that for all $k_0 \in \mathbb{N}$*

$$\alpha I \preceq \sum_{k=k_0}^{k_0+P-1} \Phi_k \Phi_k^\top \preceq \beta I.$$

As summarized in the following two propositions, if the regressor is PE, the point estimate as well as the set-membership estimate converge to the true parameter under suitable assumptions on the disturbance.

Proposition 4.17 (Convergence of $\hat{\theta}_k$). *Assume $w_k \equiv 0$. If the regressor $(D_k)_{k\in\mathbb{N}}$ is persistently exciting, the true parameter θ^* is a globally exponentially stable fixed point of the difference equation* (4.14).

Proposition 4.18 (Convergence of Θ_k). *Assume w_k for $k \in \mathbb{N}$ are realizations of independent random variables W_k with support $\mathbb{W}$. If the regressor $(D_k)_{k\in\mathbb{N}}$ is persistently exciting,*

$$\sup_{\theta_1,\theta_2\in\Theta_k} \|\theta_1 - \theta_2\| \to 0$$

with probability one as $k \to \infty$.

With a minor adaption due to the matrix notation and projection, the proof of Proposition 4.17 follows classical results, e.g., (Åström and Wittenmark, 2008, Theorem 6.4). The proof of Proposition 4.18 follows (Bai et al., 1998, Theorem 2.1) with again a minor difference due to the matrix notation and vector valued disturbance. The assumption on the disturbance could be weakened to the boundary of $\mathbb{W}$ being a subset of the support of W_k.

To make these results applicable within the proposed MPC framework, the following Lemma provides an explicit condition on the input sequence $(u_k)_{k\in\mathbb{N}}$, which is sufficient for the regressor $(D_k)_{k\geq n}$ being persistently exciting.

Lemma 4.19 (Persistence of excitation). *Suppose Assumption 4.4 is satisfied and $w_k = 0$. The regressor $(D_k^\top)_{k>n}$ is persistently exciting if the stacked input vector $([u_k^\top \; \cdots \; u_{k-n}^\top])_{k\geq n}$ is persistently exciting.*

The proof of Lemma 4.19 is given in Appendix B.3. The assumption of $w_k = 0$ is of technical nature and only required to ensure that the disturbance does not "counteract" the input. The required PE condition on the input sequence can directly be incorporated in the proposed MPC framework, e.g., by augmenting the cost function (Heirung et al., 2017), by introducing an additional constraint (Marafioti et al., 2014) or within a two-step procedure where first an optimal input is derived which is then suitably perturbed (Tanaskovic et al., 2014).

Remark 4.20. We remark that, as an alternative to persistence of excitation, the initially discussed idea of Dual Control extends to the presented robust setting when future learning is taken into account in the cost function (Veres, 1995). However, as with a stochastic system description, this leads to an intractable, non-convex optimization program, not suitable for use in model predictive control.

4.4 Numerical example and application in tracking MPC

To illustrate the advantages of the proposed MPC with online parameter identification, two numerical examples are discussed in the following. In the first example, the focus is put on stability and constraint satisfaction. Additionally, convergence of the online parameter identification under the discussed PE constraint on the input is shown. In the second example, we compare the tracking performance of the proposed adaptive MPC with a non-adaptive, robust MPC.

Example 1

Consider the second-order discrete-time linear system (4.1) with

$$
\begin{aligned}
A_0 &= \begin{bmatrix} 0.5 & 0.2 \\ -0.1 & 0.6 \end{bmatrix}, \quad B_0 = \begin{bmatrix} 0 \\ 0.5 \end{bmatrix}, \\
A_1 &= \begin{bmatrix} 0.042 & 0 \\ 0.072 & 0.03 \end{bmatrix}, \quad A_2 = \begin{bmatrix} 0.015 & 0.019 \\ 0.009 & 0.035 \end{bmatrix}, \quad A_3 = 0_{2\times 2}, \\
\{B_i\}_{i=1,2} &= 0_{2\times 1}, \quad B_3 = \begin{bmatrix} 0.040 \\ 0.054 \end{bmatrix}.
\end{aligned}
$$

The a priori uncertainty set is given by an infinity norm ball $\Theta = \{\theta \in \mathbb{R}^3 \mid \|\theta\|_\infty \leq 1\}$ and the initial point estimate by $\hat{\theta}_0 = 0$ while the true parameter is $\theta^* = [0.8\ 0.2\ -0.5]^\top$. The disturbances w_k, $k \in \mathbb{N}$ are independent realizations of a uniform random variable with support on $\mathbb{W} = \{w \in \mathbb{R}^2 \mid \|w\|_\infty \leq 0.1\}$. The system is subject to separate state and input constraints

$$
[x_k]_2 \geq -0.3, \qquad |u_k| \leq 1,
$$

which should be satisfied robustly. An additional box constraint $\|x_k\|_\infty \leq 3$ has been introduced to determine the parameter update gain μ. In the simulation, the MPC cost weights were set to $Q = \mathrm{diag}(1,1)$, $R = 1$, and the prediction horizon to

$N = 10$. For disturbance attenuation, the prestabilizing and terminal control gain was set to $K = [0.017 \ -0.41]$.

Figure 4.1 shows a closed-loop state trajectory resulting from application of Algorithm 4.12 with the initial condition $x_0 = [2\ 3]^\top$ and a random disturbance sequence. In addition, the initially predicted state tube and the state constraint $[x_k]_2 \geq -0.3$ is plotted. As proven in Theorem 4.13, the state constraint is robustly satisfied for all possible predicted as well as realized states and the state converges to a neighborhood of the origin. The state constraint is active initially, which confirms Proposition 4.8 as the state constraint is tangent to the depicted sets $\mathbb{X}_{l|k}$ for $l \geq 3$. Similarly, the input constraints are satisfied robustly with the constraint $u_k \geq -1$ being active initially.

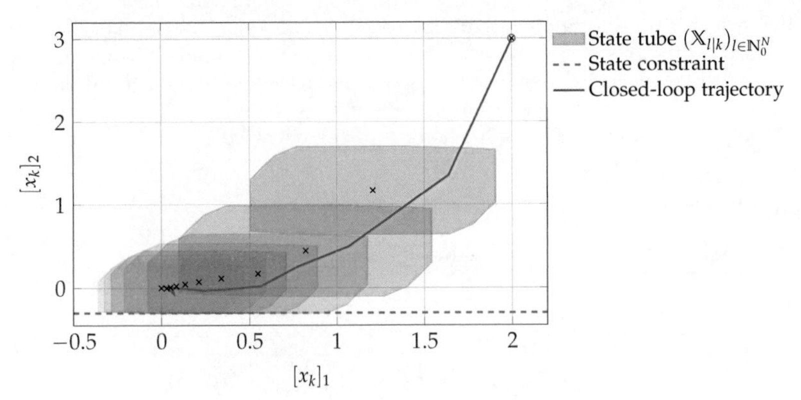

Figure 4.1. Realized closed-loop trajectory from initial condition $x_0 = [2\ 3]^\top$, predicted state tube at time $k = 0$, and constraint $[x_k]_2 \geq -0.3$.

To illustrate convergence of the parameter estimates, the discussed persistence of excitation condition has been implemented. In particular, following Lemma 4.19, the additional input constraint

$$\sum_{l=0}^{P-1} \mathbf{u}_{k-l}\mathbf{u}_{k-l}^\top \succeq \alpha I \tag{4.25}$$

with $P = n + 1$ and $\alpha = 2$ has been added to the MPC optimization.

Figure 4.2 shows the simulated closed-loop trajectory resulting from application of Algorithm 4.12 with the additional input constraint 4.25, initial condition

$x_0 = [0\ 0]^\top$ and a random disturbance sequence. As expected, the closed loop exhibits a persistently exciting regressor, with the typical cyclic state and input trajectories. Note that the center of the trajectory path is shifted from the origin to the positive orthant as the closed-loop trajectory needs to satisfy the state constraint $[x_k]_2 \geq -0.3$. The estimated parameter membership set Θ_k is shown in Figure 4.3. As predicted by Lemma 4.2 and Proposition 4.18, the set is non-increasing and converges to a singleton.

The simulations were performed in Matlab with the modeling language Yalmip to define and MOSEK to solve the optimization program (4.23). With the state tube cross sections depicted in Figure 4.1, the median solver time (with PE constraint) reported by Yalmip was 0.068s (0.10s) with a maximum of 0.095s (0.19s) and minimum of 0.05s on an Intel Core i7 with 3.4GHz. The main impact on the solver time has the shape of the tube cross sections, an increased complexity of $\mathbb{X}_0$ significantly increases the number constraints. Choosing $\mathbb{X}_0$ to be the minimal robust forward invariant set under the local control law decreases conservatism, but a significant increase in computation time has been observed in the simulations. The non-convex PE constraint 4.25 has been decomposed into two linear and one integer constraint as noted in (Marafioti et al., 2014). This led to two convex QP problems to be solved and compared in each MPC iteration instead of one non-convex problem.

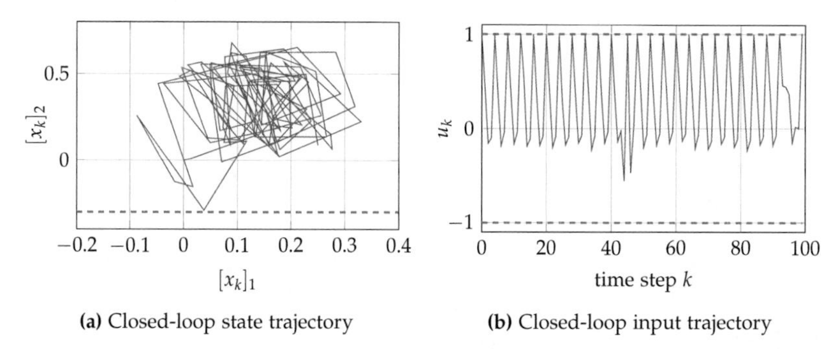

(a) Closed-loop state trajectory **(b)** Closed-loop input trajectory

Figure 4.2. Closed-loop state and input trajectory resulting from application of Algorithm 4.12 with PE input constraint (4.25). The state and input constraints are depicted as dashed lines.

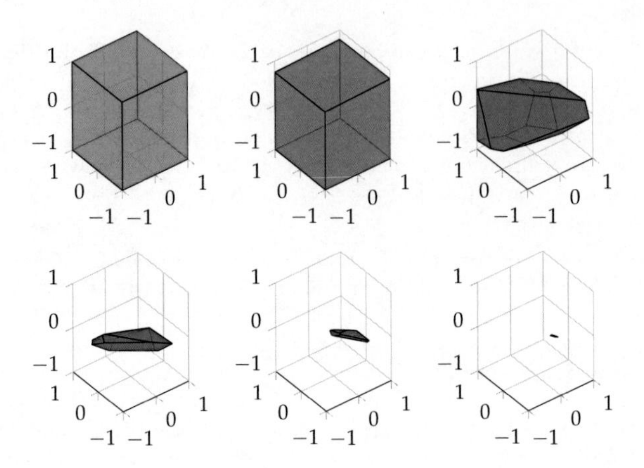

Figure 4.3. Parameter membership set at time steps $k = 0, 5, 25, 70, 120, 500$ resulting from application of Algorithm 4.12 with PE input constraint (4.25). The membership set is non-increasing and converges to a singleton containing the true parameter $\theta^* = [0.8\ 0.2\ {-0.5}]^\top$.

Example 2

To highlight the increased performance gained from the online parameter identification, we compare it with a non-adaptive tube MPC in this second example. Consider the mass spring damper system with dynamics given by

$$m\ddot{y} = -c\dot{y} - ky + k_u u + w.$$

The nominal parameters are mass $m = 1$, damping constant $c = 0.2$, spring constant $k = 1$, and input gain $k_u = 1$. The parameter uncertainty for the coefficients $\frac{c}{m}$, $\frac{k}{m}$, and $\frac{k_u}{m}$ are $\pm 20\%$ and the disturbance is bounded by $|w(t)| \leq 0.5$. The input force u is constrained to $[-5, 5]$ and the position y to $[-0.1, 1.1]$. To apply the presented MPC algorithm, a state-space representation and first-order discretization with sampling time $T_s = 0.1$ was used. The cost weights were set to $Q = \operatorname{diag}(10, 0.001)$, $R = 0.001$, and the prediction horizon to $N = 14$.

Instead of stabilizing the origin, the control task is to stabilize a setpoint which is switched between 0 and 1 at time steps $k = 10, 30, 50, 70$. Figure 4.4 shows the state constraints and reference trajectory jointly with the closed-loop response under the presented MPC algorithm with and without parameter adaption. As expected, up to time $k = 10$ the response to the first setpoint change is similar for the adaptive and robust MPC. However, the adaptive MPC becomes more aggressive in the subsequent setpoint changes as the model uncertainty is decreased. Similarly, while each transient between the desired steady states is identical for the robust MPC, the adaptive MPC shows a faster, improved convergence at each setpoint change. In particular, the presented MPC scheme with online parameter estimation is able to reach the desired steady state within 10 time steps, whereas the robust MPC does not converge to the desired steady state before the subsequent set-point change. The root-mean-square tracking error of the robust MPC is 0.20 compared to 0.14 for the adaptive MPC, i.e., 43% higher.

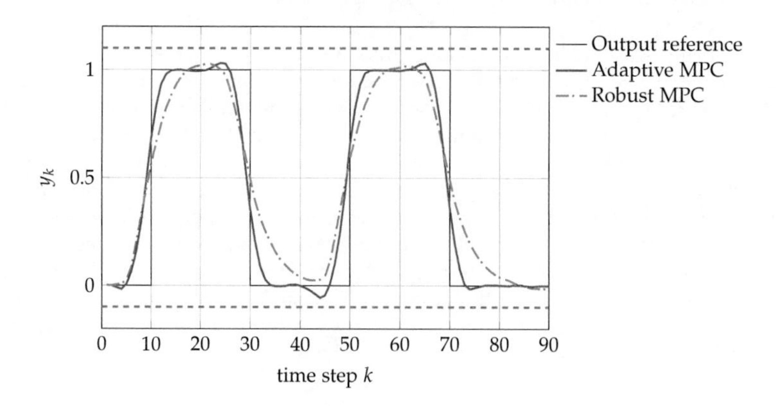

Figure 4.4. Comparison of the closed-loop trajectories resulting from application of Algorithm 4.12 and a non-adaptive robust MPC. The adaptive MPC shows an increasingly better tracking performance for the desired setpoint changes at $k = 10, 30, 50, 70$.

4.5 Discussion and summary

To conclude this chapter, we discuss the limitations of the proposed adaptive MPC algorithm and briefly summarize the obtained results.

Limitations of the proposed approach

Conceptually, the main limitation is the conservatism due to the robust, set-based approach. Compared to the classical adaptive control literature, the assumptions are rather restrictive and allow to implement a robust MPC algorithm. Yet, this is to be expected if state constraints are to be satisfied robustly. The main advantage over a robust MPC is a significant performance increase, which can be expected from the proposed adaptive approach as illustrated in the second example. An ad-hoc solution to overcome these restrictions is the use of soft-constraints in the form of penalty functions. A rigorous stochastic formulation based upon the presented results is currently under investigation and results will be presented in (Andersson, 2018).

On a computational level, the main limitation stems from the use of polytopes, which generally scale badly with the system dimension. In particular, the complexity of the tube cross section $\mathbb{X}_0$ significantly impacts the number of constraints and thereby the required time to solve the online optimization. Similarly the complexity of the set-membership estimation increases with an increasing dimension of the uncertain parameter vector. Yet, we remark that through the use of a state space model and prediction tube, the dimension of the employed polytopes and thereby number of constraints is generally much smaller than if an FIR model is employed and the parameters and complete trajectories are bounded directly as proposed in (Tanaskovic et al., 2014). Furthermore, the presented approach easily extends to using ellipsoids instead of polytopes which increases the conservatism but scales nicely with the system dimension. Note that the resulting optimization program then becomes a second order cone program instead of a linearly constrained quadratic program.

Finally, from an implementation point of view, the computation of the terminal set and the non-standard constraints increase the required engineering effort of implementing the proposed adaptive MPC compared to a nominal MPC. Similarly, the bounds for the uncertain parameters and disturbance should be chosen with care to include all possible realizations without increasing the conservatism too much. Due to the set-membership estimation, an explicit handling of outliers and a possible sanity check is required in real world applications.

Summary

In contrast to the previously discussed setup of independent and identically distributed disturbances, in this chapter, we introduced the problem of constrained control of systems with constant or slowly time-varying uncertainties. For the class of linear systems, we provided a computationally tractable framework based upon set-membership system identification instead of a probabilistic description with recursive Bayesian parameter estimation.

We first introduced the concept of indirect adaptive MPC and briefly discussed the requirements for stability and constraint satisfaction. This eventually led to a set-membership parameter estimation to derive bounds on the state and input predictions in combination with a projected Least Mean Squares filter to achieve a finite gain from the disturbance to the prediction error. As a sufficient condition to derive closed-loop stability, we gave an upper bound on the parameter update gain μ which directly depends on the constraint set. Thereby we naturally recovered the common recommendation of an inverse correlation of the parameter update gain and the maximum absolute value of the state and input values.

Using strong duality in linear programming, we reformulated the conditions for constraint satisfaction under all possible disturbance and uncertainty realizations as linear constraints. One key enabler of this reformulation was the use of homothetic tubes to bound the state and input predictions, in particular the explicit characterization of the vertices and half-spaces of the predicted polytopes $\mathbb{X}_{l|k}$ which contain all possible state realizations.

Finally, this led to the MPC optimization program being a convex, linearly constrained quadratic program, which has been proven to be recursively feasible. Furthermore we proved a finite ℓ_2 gain of the closed loop as well as robust constraint satisfaction. Since the cost function does not explicitly incorporate the benefit of future learning, a sufficient condition for persistence of excitation has been derived under which convergence of the parameter estimates could be established.

To the best of our knowledge, this is the first indirect adaptive MPC scheme in a modern state space formulation which makes full use of the online parameter estimation and where the online optimization could be reduced to a computationally tractable convex program. This makes the method readily applicable and provides a solid adaptive MPC framework which can be easily combined with further results tailored to specific control objectives, e.g., tracking or output feedback MPC. While we only considered static uncertainties, the results directly extend to time-varying parameters as presented in (Lorenzen et al., 2018).

Chapter 5

Stochastic MPC without terminal constraints

One of the main sources of conservatism and computational difficulty in the foregoing chapters was the terminal constraint. On the one hand, given the stochastic disturbance model, the uncertainty in the predicted states and inputs increases with the horizon length. This makes terminal constraints even more restrictive than in the nominal case, since the terminal set should contain the last predicted state with high probability. On the other hand, it can be non-trivial to find a suitable terminal controller and terminal set that is robust forward invariant under the local dynamics as required by the Assumption 2.6. This led to an increasing computational complexity in Chapter 3 and the non-standard Assumption 4.3 in Chapter 4. In contrast, omitting the terminal constraints and cost allows for a significantly easier, straight-forward practical implementation. These arguments motivate the development of theory for and computational approaches to stabilizing stochastic MPC without terminal constraints and without a terminal cost.

In this chapter, we first derive a *conceptual* stochastic MPC algorithm without terminal constraints or terminal cost and analyze the convergence of the closed loop, which brings us full circle with regard to Chapter 2. Unfortunately, as the usual stability proofs strongly depend on the principle of optimality, they only extend to the conceptual algorithm, but not to computational tractable approximations as presented in Chapters 3 and 4. Hence, based upon the conceptual results, the assumptions are strengthened to derive *computational approaches* for i) optimization over parametrized feedback policies and ii) the relaxation of chance constraints that were introduced in Chapter 3. While the main results are presented for the general case of non-linear systems, we explicitly discuss simplifications and computational approaches for the class of linear systems.

The remainder is structured as follows. After introducing the problem setup and basic assumptions in Section 5.1, we present theoretical results for the conceptual, ideal stochastic MPC algorithm in Section 5.2. Sufficient conditions to derive stability for computational tractable algorithms are derived and discussed in Section 5.3. Finally, Section 5.4 provides a brief discussion and summary of the obtained results.

This chapter is based on (Lorenzen et al., 2017d,e).

5.1 Problem setup and conceptual SMPC algorithm

The basic problem setup as well as the assumptions on the system model and constraints are similar to Section 2.2. For completeness, they are again briefly summarized below. The main differences are in the following SMPC finite horizon optimal control problem and the related assumptions.

Let the stochastic discrete time system be modeled by the realization of a controlled Markov process

$$x_{k+1} = f(x_k, u_k, w_k) \tag{5.1}$$

with state $x_k \in \mathbb{R}^n$, controlled input $u_k \in \mathbb{R}^m$, and exogenous disturbance input $w_k \in \mathbb{R}^{m_w}$ satisfying Assumption 2.4. Given the state x_k at time k, the predicted, future states are given by

$$x_{l+1|k} = f(x_{l|k}, u_{l|k}(x_{l|k}), W_{l+k}) \qquad x_{0|k} \overset{a.s.}{=} x_k,\ W_{l+k} \sim \mathbb{P}_w \tag{5.2}$$

where $x_{l|k}$ are random variables and $u_{l|k} : \mathbb{R}^n \to \mathbb{R}^m$ are Borel measurable functions.

The system is subject to hard constraints on the input

$$u_{l|k}(x_{l|k}) \in \mathbb{U} \quad \mathbb{P}_k \ a.s. \tag{5.3}$$

with $\mathbb{U} = \{u \in \mathbb{R}^m \mid c_u(u) \leq 0\}$, $c_u : \mathbb{R}^m \to \mathbb{R}$ and probabilistic constraints on the state

$$\mathbb{P}\{x_{l+1|k} \in \mathbb{X} \mid x_{l|k}\} \geq 1 - \varepsilon \quad \mathbb{P}_k \ a.s. \tag{5.4}$$

with $\mathbb{X} = \{x \in \mathbb{R}^n \mid c_x(x) \leq 0\}$, $c_x : \mathbb{R}^n \to \mathbb{R}$. The performance of the control system is measured by a running cost $\ell : \mathbb{R}^n \times \mathbb{R}^m \to \mathbb{R}_{\geq 0}$, which penalizes the distance to the desired setpoint $\bar{x} = 0$ and input $\bar{u} = 0$.

Assumption 5.1 (Continuity). *The stage cost ℓ as well as constraint functions c_u and c_x are continuous.*

In contrast to the previous chapters, in the following, the stochastic MPC control law is based upon a finite horizon optimal control problem that does not incorporate a terminal cost and constraint, i.e.,

$$V_N(x_k) = \min_{\mathbf{u}_{N|k}} \mathbb{E}_k \left\{ \sum_{l=0}^{N-1} \ell(x_{l|k}, u_{l|k}(x_{l|k})) \right\} \tag{5.5a}$$

$$\text{s.t. } x_{l+1|k} = f(x_{l|k}, u_{l|k}(x_{l|k}), W_{l+k}), \qquad x_{0|k} \overset{a.s.}{=} x_k \tag{5.5b}$$

$$\mathbb{P}\{x_{l+1|k} \in \mathbb{X} \mid x_{l|k}\} \geq 1 - \varepsilon \quad \mathbb{P}_k \text{ a.s.} \qquad \forall l \in \mathbb{N}_0^{N-1} \tag{5.5c}$$

$$u_{l|k}(x_{l|k}) \in \mathbb{U} \quad \mathbb{P}_k \text{ a.s.} \qquad \forall l \in \mathbb{N}_0^{N-1}. \tag{5.5d}$$

The stochastic MPC algorithm without terminal constraints is then equivalent to the basic MPC Algorithm 2.1 with the online optimization replaced by (5.5). Assuming a minimizer $\mathbf{u}^*_{N|k}$ exists, the SMPC control law is defined by $\kappa(x_k) = u^*_{0|k}(x_k)$.

In the following, we introduce and discuss assumptions which are sufficient to prove stability of the closed loop and thereby replace the basic stability Assumption 2.6 of Chapter 2.

Assumption 5.2 (Viability). *There exists a robust controlled invariant set $\mathbb{X}_\infty$ such that for all $x_k \in \mathbb{X}_\infty$, $N \in \mathbb{N}$ the optimal value $V_N(x_k)$ is attained and for all $w \in \mathbb{W}$ it holds $f(x_k, \kappa(x_k), w) \in \mathbb{X}_\infty$.*

In the previous chapters, we were able to prove recursive feasibility by choosing a suitable terminal set and control law. In contrast, in this chapter, Assumption 5.2 is introduced to ensure feasibility of the online optimization and thereby helps to have a well defined control law and avoid unnecessary technicalities in the following presentations. For computational methods to verify Assumption 5.2 we refer the interested reader to (Blanchini and Miani, 2015) for linear systems and (Aubin et al., 2011) as well as (Grüne and Pannek, 2017, Chapter 8) for a more general treatment of the topic.

Assumption 5.3 (Cost function).

i) Let $\mathbf{u}^0_{N|k}$ be the optimal solution to the MPC optimization (5.5) with horizon length N and initial condition $x_k = 0$. Let $\mathbf{x}^0_{N|k}$ be the corresponding predicted state trajectory and denote the corresponding stage cost $\ell_l^N = \mathbb{E}_k\{\ell(x^0_{l|k}, u^0_{l|k}(x^0_{l|k}))\}$. Let $\mathbf{u}^*_{N|k}$ and $\mathbf{x}^*_{N|k}$ be the optimal predicted input functions and state trajectories for any $x_k \in \mathbb{X}_\infty$. The optimal cost satisfies $\ell_0^N = 0$, $\mathbb{E}_k\{\ell(x^*_{l|k}, u^*_{l|k}(x^*_{l|k}))\} \geq \ell_l^N$, and $\ell^N_{N-l} \geq \ell^{N'}_{N'-l}$ for $N \geq N'$.

ii) There exists a continuous, positive definite, radially unbounded function $\sigma : \mathbb{R}^n \to \mathbb{R}_{\geq 0}$ and $\bar{a} \in \mathbb{R}_{\geq 0}$ such that $V_N(x) \leq \bar{a}\sigma(x) + \sum_{l=0}^{N-1} \ell_l^N$ and $\ell(x, u) \geq \sigma(x)$ for all $u \in \mathbb{U}$.

As discussed and motivated in Chapter 2, we do not assume a cost which is positive definite with respect to a reachable steady-state distribution but with respect to the desired setpoint $\bar{x} = 0$ and input $\bar{u} = 0$. Assumption 5.3 i) imposes that the expected running cost is minimal for an initial condition at the desired setpoint and non-decreasing with the prediction time. Assumption 5.3 ii) imposes an upper bound on the optimal cost V_N and radial unboundedness of the running cost ℓ. The assumptions are similar to the assumptions taken in (Grimm et al., 2005) albeit adapted to the stochastic setting and the possibly non-zero steady-state cost.

As shown in Section 5.3, suitable bounds can be derived constructively for the class of linear systems with additive disturbance. For arbitrary nonlinear systems, a general systematic approach is not available. Yet, the following controllability assumption, which extends the one for nominal MPC introduced in (Grüne and Pannek, 2017), could be employed to verify Assumption 5.3 ii).

Definition 5.1 (Stochastic controllability). *System (5.1) is stochastically controllable with respect to ℓ with rate $\beta \in \mathcal{KL}$, if for each $x_k \in \mathbb{X}_\infty$ and $N \in \mathbb{N}$ there exists an admissible control $\tilde{u}_{l|k}(\cdot) \in \mathbb{U}$ as well as state sequence $\tilde{x}_{l|k}$ satisfying $\mathbb{P}\{\tilde{x}_{l+1|k} \in \mathbb{X} \mid \tilde{x}_{l|k}\} \geq 1 - \varepsilon$ and*

$$\mathbb{E}_k \left\{ \ell(\tilde{x}_{l|k}, \tilde{u}_{l|k}(\tilde{x}_{l|k})) \right\} \leq \beta(\sigma(x_k), l) + \ell_l^N$$

for all $l \in \{1, \dots, N-1\}$ and $\tilde{x}_{l+1|k} = f(\tilde{x}_{l|k}, \tilde{u}_{l|k}(\tilde{x}_{l|k}), W_{l+k})$ with $\tilde{x}_{0|k} \overset{a.s.}{=} x_k$.

An upper bound $\bar{\alpha}$ satisfying Assumption 5.3 is directly available if a system is stochastically controllable with respect to ℓ and a rate β that is linear in its first argument and belongs to the class of summable $\mathcal{KL}$-functions, i.e., $\sum_{k=0}^{\infty} \beta(r, k)$ is finite for all $r \geq 0$ (Grüne and Pannek, 2017). Along the lines of the following derivations, this property could be used to strengthen the results presented in this chapter, in particular to derive a lower sufficient bound on the length of the prediction horizon. Yet, we focus on the more general Assumption 5.3 ii) in order to highlight the fundamental differences to nominal SMPC, which are relevant when approximations to (5.5) are employed. For the same reason we also do not focus on a tight derivation of the minimal stabilizing horizon under the given or possibly stronger assumptions.

Finally, note that more general upper and lower bounds $\underline{\alpha}_l, \bar{\alpha} \in \mathcal{K}_\infty$ such that $V_N(x) \leq \bar{\alpha}(\sigma(x)) + \sum_{l=0}^{N-1} \ell_l$ and $\ell(x, u) \geq \underline{\alpha}_l(\sigma(x))$ for all $u \in \mathbb{U}$ could be assumed instead of Assumption 5.3 ii), but would lead to weaker stability results.

In the following, our main objective is to discuss closed-loop properties when the system (5.1) is controlled with the MPC control law $\kappa(x_k) = u_{0|k}^*(x_k)$ and computationally tractable approximations thereof.

5.2 Closed-loop properties

Analogous to the development from Chapter 2 to Chapter 3 and 4, we first analyze the previously introduced idealized setup where the predicted states are random variables and the optimization is performed over measurable functions $u_{l|k} : \mathbb{R}^n \to \mathbb{R}^m$. The proof of the following theorem is constructive, leading to explicit bounds on the introduced constants c_1, λ, and the prediction horizon length N.

Theorem 5.2 (Unconstrained SMPC closed-loop properties). *Consider the stochastic MPC algorithm based on optimization* (5.5). *Suppose Assumption 5.1-5.3 are satisfied and* $x_0 \in \mathbb{X}_\infty$.

For all $k \in \mathbb{N}$, *the state and input trajectories of the closed-loop system* (5.1) *with SMPC feedback law* $\kappa(x_k) = u^*_{0|k}(x_k)$ *satisfy the input constraint* $u_k \in \mathbb{U}$ *almost surely and the state constraint* $x_k \in \mathbb{X}$ *with at least probability* $1 - \varepsilon$. *There exists* $c_1 \in \mathbb{R}_{\geq 0}$ *such that*

$$\mathbb{E}_k \left\{ V_N(x_{k+1}) \right\} - V_N(x_k) \leq -\ell(x_k, u_k) + \frac{\bar{a}^2}{N-1}\sigma(x) + c_1. \tag{5.6}$$

Furthermore, there exists $N \in \mathbb{N}$ *such that for all* $x_k \in \mathbb{X}_\infty$ *it holds*

$$\mathbb{E}_k \left\{ V_N^0(x_{k+1}) \right\} \leq \lambda V_N^0(x_k) + c_1 \tag{5.7}$$

with $\lambda \in (0,1)$ *and* $V_N^0(x) = V_N(x) - \sum_{l=0}^{N-1} \ell_l^N$.

Under the stated assumptions, Theorem 5.2 guarantees qualitatively the same closed-loop properties as if terminal constraints and a terminal cost was used. In particular, there exists a sufficiently large optimization horizon N such that the convergence results and the limiting behavior stated in Corollaries 2.5, 2.6, and 2.7 hold.

To prove the result, recall the stochastic dynamic programming principle, which will be exploited in the following. With $V_0 \equiv 0$, for $N \geq 0$ and $x \in \mathbb{X}_\infty$ it holds

$$\begin{aligned} V_N(x) = \min_u\ & \ell(x,u) + \int_{\mathbb{W}} V_{N-1}(f(x,u,w))\mathbb{P}(\mathrm{d}w) \\ \text{s.t. }\ & u \in \mathbb{U} \\ & \mathbb{P}\{f(x,u,W) \in \mathbb{X}\} \geq 1 - \varepsilon. \end{aligned} \tag{5.8}$$

The stability result for the conceptual SMPC algorithm then follows similar arguments as in nominal MPC presented in (Grimm et al., 2005) albeit adapted to the stochastic system and in particular taking into account that $\bar{x}$, $\bar{u}$ is not a steady state.

If the system satisfies Definition 5.1, arguments paralleling (Grüne and Pannek, 2017) can be made to derive a tighter bound for a sufficiently large optimization horizon N. The proof highlights structural differences between establishing stability for the conceptual algorithm and for the following SMPC schemes, which are based upon approximations of the finite horizon optimal control problem.

Proof of Theorem 5.2. Analogous to Proposition 2.4, constraint satisfaction follows directly from the definition of the constraints and the optimization program being feasible. For convergence we consider the expected difference in the optimal value function from time k to $k+1$:

$$
\begin{aligned}
&\mathbb{E}_k\left\{V_N(x_{k+1})\right\} - V_N(x_k) \\
&\quad = \mathbb{E}_k\left\{\sum_{l=0}^{N-j-1} \mathbb{E}_{k+1}\left\{\ell(x^*_{l|k+1}, u^*_{l|k+1}(x^*_{l|k+1}))\right\} + V_j(x^*_{N-j|k+1})\right\} - V_N(x_k) \\
&\quad \leq -\sum_{l=0}^{N-1} \mathbb{E}_k\left\{\ell(x^*_{l|k}, u^*_{l|k}(x^*_{l|k}))\right\} + \sum_{l=1}^{N-j} \mathbb{E}_k\left\{\ell(x^*_{l|k}, u^*_{l|k}(x^*_{l|k}))\right\} \\
&\qquad + \mathbb{E}_k\left\{V_j(x^*_{N-j+1|k})\right\} + \sum_{l=N-j+1}^{N-1} \ell_l^N - \sum_{l=1}^{j-1} \ell_l^j \\
&\quad \leq -\ell(x_k, u_k) + \mathbb{E}_k\left\{V_j(x^*_{N-j+1|k})\right\} - \sum_{l=1}^{j-1} \ell_l^j \\
&\quad \leq -\ell(x_k, u_k) + \bar{a}\, \mathbb{E}_k\left\{\sigma(x^*_{N-j+1|k})\right\}.
\end{aligned}
\tag{5.9}
$$

The first equality follows from the dynamic programming principle (5.8) applied j times. The first inequalities, which holds for arbitrary $j \leq N$, is based upon employing $u_{l|k+1} \equiv u^*_{l+1|k}$ as feasible but possibly suboptimal solution for $l \leq N-j$ and adding the non-negative term $\sum_{l=N-j+1}^{N-1} \ell_l^N - \sum_{l=1}^{j-1} \ell_l^j$. Assumption 5.3 is employed again to obtain the second and third inequality.

By Assumption 5.3 ii) on the upper limit on V_N we have

$$
V_N(x_k) = \sum_{l=1}^{N-1} \mathbb{E}_k\left\{\ell(x^*_{l|k}, u^*_{l|k}(x^*_{l|k}))\right\} \leq \bar{a}\sigma(x_k) + \sum_{l=0}^{N-1} \ell_l^N
$$

and hence there exists $l' \in \{1, \ldots, N-1\}$ such that

$$
\mathbb{E}_k\left\{\sigma(x^*_{l'|k})\right\} \leq \mathbb{E}_k\left\{\ell(x^*_{l'|k}, u^*_{l'|k}(x^*_{l'|k}))\right\} \leq \frac{\bar{a}}{N-1}\sigma(x_k) + \bar{\ell} \tag{5.10}
$$

with $\bar{\ell} = \frac{1}{N-1}\sum_{l=0}^{N-1} \ell_l^N$. Choosing $j = N - l' + 1$, the last inequality inserted in (5.9) leads to

$$\mathbb{E}_k \left\{V_N(x_{k+1})\right\} - V_N(x_k) \leq -\ell(x_k, u_k) + \frac{\bar{a}^2}{N-1}\sigma(x_k) + \bar{a}\bar{\ell} \tag{5.11}$$

which proves the first part of the theorem with $c_1 = \bar{a}\bar{\ell}$. For the second part note that for $N \geq \bar{a}^2 + 1$ it holds

$$\begin{aligned} -\ell(x_k, u_k) + \frac{\bar{a}^2}{N-1}\sigma(x_k) + \bar{a}\bar{\ell} &\leq \left(\frac{\bar{a}^2}{N-1} - 1\right)\sigma(x_k) + \bar{a}\bar{\ell} \\ &\leq \left(\frac{\bar{a}}{N-1} - \frac{1}{\bar{a}}\right)V_N^0(x_k) + \bar{a}\bar{\ell} \end{aligned}$$

which proves the claim with $\lambda = \frac{\bar{a}^2+(\bar{a}-1)(N-1)}{\bar{a}(N-1)}$. Note that $\lambda < 1$ for $N > \bar{a}^2 + 1$. ∎

The following example, which will be used throughout this chapter, illustrates the result and touches upon the conservatism, which is due to using Assumption 5.3 and a cost which is not positive definite with respect to a steady state.

Example 5.3. Consider a discrete time integrator described by $x_{k+1} = x_k + u_k + w_k$ with $W_k \sim \mathbb{P}_w$, $\mathbb{E}\{W_k\} = 0$, $\mathbb{E}\{W_k^2\} = \sigma_w^2$, and running cost $\ell(x,u) = x^2 + u^2$. Assumption 5.1 and 5.2 are (trivially) satisfied. Let K_l be the finite horizon LQ optimal feedback gains for horizon length N, define $e_{l+1} = (1 + K_l)e_l + W_l$ with $e_0 \overset{a.s.}{=} 0$, and let P be the solution of the discrete time algebraic Ricatti equation. Assumption 5.3 is satisfied with $\ell_l^N = (1 + K_l^2)\mathbb{E}\{e_l^2\}$, $\sigma(x) = x^2$, and $\bar{a} = P$. By Theorem 5.2 it holds

$$\mathbb{E}_k\{V_N(x_{k+1})\} - V_N(x_k) \leq -(x_k^2 + u_k^2) + \frac{P^2}{N-1}x_k^2 + c_1$$

with $c_1 = P\bar{\ell}$ and for $N > P^2 + 1$ by Corollary 2.5 it holds

$$\lim_{t\to\infty} \frac{1}{t}\sum_{k=0}^{t-1} \mathbb{E}\left\{x_k^2\right\} \leq \frac{(N-1)P}{(N-1) - P^2}\bar{\ell}$$

where the right hand side is determined by summing up (5.6) over k and taking iterated expectations.

For $N \to \infty$ it holds $\bar{\ell} = P\sigma_w^2$ and hence $\lim_{t\to\infty} \frac{1}{t}\sum_{k=0}^{t-1} \mathbb{E}\left\{x_k^2\right\} \leq P^2\sigma_w^2$ instead of the true, tight bound $P\sigma_w^2$. The conservativeness comes from "not knowing" l' or j, respectively: For $N \to \infty$, both, (5.9) and (5.10) are tight in the example, but for different values of l' and j, respectively. Choosing $j = N$ in (5.9), the last inequality leads to $\mathbb{E}_k\{V_N(x_{1|k})\} - \sum_{l=1}^{N-1} \ell_l^N \leq P\mathbb{E}_k\{x_{1|k}^2\}$ which is tight for $N \to \infty$. Choosing $j = 2$, i.e., $l' = N - 1$ in (5.10), we have $\mathbb{E}_k\{x_{N-1|k}^2\} \leq \frac{Px_k^2}{N-1} + \bar{\ell}$, which again is tight for $N \to \infty$. Similar to (Grüne and Pannek, 2017), the controllability assumption given in Definition 5.1 or a cost which is positive definite with respect to the optimal steady state *distribution* could be used to tighten the result.

5.3 Stability under approximations and relaxations

As discussed in Chapter 2, the above introduced stochastic MPC algorithm is in general not tractable and approximations as introduced in Chapter 3 and 4 are necessary to derive computational approaches. Thereby, the focus has been on deriving suitable, computationally tractable relaxations while still being able to prove recursive feasibility and convergence of the closed loop. In the following, we discuss the previously presented computational approaches for stochastic MPC without terminal constraint. Yet, as the above presented stability proof strongly relies on the principle of optimality, which holds only for an exact solution of the finite horizon stochastic optimal control problem (5.5), the argument does not extend trivially but necessitates further analysis and assumptions.

Stability with input parameterization

Using input parameterizations, the dynamic programming equation (5.8) does not apply and, hence, inequality (5.9) is invalidated. In the following, we seek to replace (5.8) in order to recover the convergence results of Theorem 5.2.

In the following, let $u_{l|k}$ be parameterized input functions where the new stochastic MPC optimization variable $p_{l|k} \in \mathbb{R}^{m_p}$ is mapped to measurable functions $u_{l|k}^p : \mathbb{R}^n \to \mathbb{R}^m$. Examples include $u_{l|k}^p(x) = Kx + v_{l|k}$ with given prestabilizing feedback K and optimization variables $p_{l|k} = v_{l|k} \in \mathbb{R}^m$ as used for linear systems in Chapter 3 or more general parameterization $u_{l|k}^p(x) = \sum_{i=1}^{m_p} [p_{l|k}]_i u_i(x)$ with given basis functions u_i as proposed in (Skaf and Boyd, 2009). With this, consider the finite horizon optimal control program (5.5) but optimization over the real-valued parameters $\mathbf{p}_{N|k} = (p_{l|k})_{l \in \mathbb{N}_0^{N-1}}$ instead of functions $\mathbf{u}_{N|k}$, i.e.,

$$\begin{aligned} \tilde{V}_N(x_k) = \min_{\mathbf{p}_{N|k}} \ & \mathbb{E}_k \left\{ \sum_{l=0}^{N-1} \ell(x_{l|k}, u_{l|k}(x_{l|k})) \right\} \\ \text{s.t.} \ & u_{l|k} = u^p_{l|k} \\ & (5.2), (5.3), (5.4) \qquad \forall l \in \mathbb{N}_0^{N-1} \end{aligned} \tag{5.12}$$

and denote its minimizer by $\mathbf{p}^*_{N|k}$. Furthermore, we introduce the end piece of length $j \leq N$ of the parameterized stochastic MPC optimization (5.12) and denote its optimal value function by $\tilde{V}_{N,j}$

$$\begin{aligned} \tilde{V}_{N,j}(x_{N-j|k}) = \min_{p_{N-j|k}, \dots, p_{N-1|k}} \ & \mathbb{E}_k \left\{ \sum_{l=N-j}^{N-1} \ell(x_{l|k}, u_{l|k}(x_k)) \right\} \\ \text{s.t.} \ & u_{l|k} = u^p_{l|k} \\ & (5.2), (5.3), (5.4) \qquad \forall l \in \mathbb{N}_{N-j}^{N-1}. \end{aligned} \tag{5.13}$$

Note that $x_{N-j|k}$ is a random variable and the expected value is taken over the joint probability distribution of $x_{N-j|k}$ and $W_{k+N-j}, \dots, W_{k+N-2}$.

Using $\tilde{V}_{N,j}$, the dynamic programming principle (5.8) can be restated as follows.

Lemma 5.4. *With $\tilde{V}_{N,0} \equiv 0$, for $j \in N_1^N$ and $x \in \mathbb{X}_\infty$ it holds*

$$\begin{aligned} \tilde{V}_{N,j}(x) = \min_{p} \ & \ell(x, u^p(x)) + \tilde{V}_{N,j-1}(f(x, u^p(x), W)) \\ \text{s.t.} \ & u^p(x) \in \mathbb{U} \\ & \mathbb{P}\{f(x, u^p(x), W) \in \mathbb{X}\} \geq 1 - \varepsilon. \end{aligned} \tag{5.14}$$

The equality follows by definition of $\tilde{V}_{N,j}$ and since $\ell(x, u^p)$ is finite and does not depend on $p_{l|k}$ for $l \in \mathbb{N}_{N-j+1}^{N-1}$. From Lemma 5.4 it can be easily seen that with parameterized feedback, (5.8) holds only with an inequality instead of the equality, in particular

$$\begin{aligned} \tilde{V}_N(x) \geq \min_{p} \ & \ell(x, u^p(x)) + \int_{\mathbb{W}} \tilde{V}_{N-1}(f(x, u^p(x), w)) \mathbb{P}(\mathrm{d}w) \\ \text{s.t.} \ & u^p(x) \in \mathbb{U} \\ & \mathbb{P}\{f(x, u^p(x), W) \in \mathbb{X}\} \geq 1 - \varepsilon. \end{aligned}$$

The inequality follows from (5.14) by noting that $\tilde{V}_{N,j}(x_{N-j|k}) \geq \mathbb{E}_k\{\tilde{V}_j(x_{N-j+k})\}$ since the solution to (5.13) is a feasible but possibly suboptimal solution to the finite horizon problem of length j starting at the *realized* state x_{N-j+k}. Taking expectations yields the result.

In order to use Lemma 5.4 in the stability proof, we strengthen Assumption 5.3 on the upper bound on the cost. In particular, an upper bound on $\tilde{V}_{N,j}$ is assumed instead of on $\tilde{V}_N$. Furthermore, an additional parameter $d \in \mathbb{R}_{\geq 0}$ is necessary due to possibly suboptimal parameterizations.

Assumption 5.4 (Cost function, SMPC with input parameterization).

i) Let $\mathbf{p}^0_{N|k}$ be the optimal solution to the parameterized MPC optimization (5.12) with horizon length N and initial condition $x_k = 0$. Let $\mathbf{x}^0_{N|k}$ be the corresponding predicted state trajectory and denote the stage cost $\ell_l^N = \mathbb{E}_k\{\ell(x^0_{l|k}, u^{p^0}_{l|k})\}$. Let $\mathbf{p}^*_{N|k}$ and $\mathbf{x}^*_{N|k}$ be the optimal solution and predicted state trajectory for any $x_k \in \mathbb{X}_\infty$. The optimal cost satisfies $\ell_0^N = 0$, $\mathbb{E}_k\{\ell(x^*_{l|k}, u^{p^*}_{l|k})\} \geq \ell_l^N$, and $\ell^N_{N-l} \geq \ell^{N'}_{N'-l}$ for $N \geq N'$.

ii) For $x_k \in \mathbb{X}_\infty$, $N \in \mathbb{N}_{>0}$ let $\mathbf{x}_{N|k}$ be the optimal predicted state trajectory. There exists a continuous, positive definite, radially unbounded function $\sigma : \mathbb{R}^n \to \mathbb{R}_{\geq 0}$ and scalars $\bar{a}, d \in \mathbb{R}_{\geq 0}$ such that

$$\tilde{V}_{N,j}(x_{N-j|k}) \leq \bar{a}\mathbb{E}_k\{\sigma(x_{N-j|k})\} + \sum_{l=0}^{j-1} \ell_l^j + d$$

for all $j \in \mathbb{N}_1^N$ and $\ell(x,u) \geq \sigma(x)$ for all $u \in \mathbb{U}$.

Under the strengthened Assumption 5.4, stability results analogous to Theorem 5.2 can be obtained for the approximate SMPC optimization with parameterized input functions.

Proposition 5.5 (Stability under input parameterization). *Consider the stochastic MPC algorithm based on optimization (5.12). Suppose Assumption 5.1, 5.2, and 5.4 are satisfied and $x_0 \in \mathbb{X}_\infty$.*

For all $k \in \mathbb{N}$, the state and input trajectories of the closed-loop system (5.1) with SMPC feedback law $\kappa(x_k) = u^{p^*}_{0|k}(x_k)$ satisfy the input constraint $u_k \in \mathbb{U}$ almost surely and the state constraint $x_k \in \mathbb{X}$ with at least probability $1 - \varepsilon$. There exists $c_1 \in \mathbb{R}_{\geq 0}$ such that

$$\mathbb{E}_k\left\{\tilde{V}_N(x_{k+1})\right\} - \tilde{V}_N(x_k) \leq -\ell(x_k, u_k) + \frac{\bar{a}^2}{N-1}\sigma(x_k) + c_1. \tag{5.15}$$

Furthermore, there exists $N \in \mathbb{N}$ such that for all $x_0 \in \mathbb{X}_\infty$ it holds

$$\mathbb{E}_k\left\{\tilde{V}^0_N(x_{k+1})\right\} \leq \lambda \tilde{V}^0_N(x_k) + c_1 \tag{5.16}$$

with $\lambda \in (0,1)$ *and* $\tilde{V}_N^0(x) = \tilde{V}_N(x) - \sum_{l=0}^{N-1} \ell_l^N$.

The proof follows analogously to the proof of Theorem 5.2 by using Lemma 5.4 instead of the dynamic programming equation (5.8), i.e., substituting $V_j(x_{N-j+1|k})$ with $\tilde{V}_{N+1,j}(x_{N-j+1|k})$ in (5.9) and applying Assumption 5.4 ii). Similarly, Corollaries 2.5, 2.6, and 2.7 hold under the assumptions given in Theorem 5.5.

Example 5.6 (Example 5.3 cont'd)**.** Consider Example 5.3 but optimization over parameterized inputs $u^p(x) = Kx + v$ with decision variable $p = v \in \mathbb{R}$ and a stabilizing, but not necessarily optimal, feedback gain K, e.g., $K = -0.5$. With $A_{cl} = A + BK$, let P satisfy the Lyapunov equation $A_{cl}^\top P A_{cl} - P + Q + K^\top R K = -0.75P + 1.25 = 0$. For $x_k = 0$ the optimal solution of (5.12) is $v_{l|k}^0 = 0$ and hence Assumption 5.4 is satisfied with $\sigma(x) = x^2$, $d = 0$, and $\ell_l^N = (1 + K^2)\mathbb{E}\{e_l^2\}$ where $e_l = \sum_{k=0}^{l-1}(1+K)^k W_k$. Furthermore it holds

$$\tilde{V}_{N,j}(x_{N-j|k}) \leq \bar{a}\mathbb{E}\left\{\sigma(x_{N-j|k})\right\} + \sum_{l=0}^{j-1} \ell_l^N = P\mathbb{E}\left\{x_{N-j|k}^2\right\} + \sum_{l=0}^{j-1} \ell_l^N$$

which follows from taking the suboptimal candidate solution $v_{l|k} = 0$ in (5.13), leading to $\bar{a} = P$.

Computational approaches for linear systems

While in principle we could recover convergence results for computational approaches to stochastic MPC without terminal constraints using parameterized input policies, similar to Assumption 5.3ii), there is no general, systematic procedure to verify the strengthened Assumption 5.4ii) for arbitrary nonlinear systems. Yet, for the class of linear systems, using the same input parameterization as in Chapter 3 and 4, in the following we discuss a constructive approach to verify Assumption 5.4 systematically.

Consider a linear, time-invariant system with additive disturbance, i.e.,

$$x_{k+1} = f(x_k, u_k, w_k) = Ax_k + Bu_k + B_w w_k. \tag{5.17}$$

Furthermore, let the constraints (5.3), (5.4) be defined by linear inequalities and single chance constraints on the state

$$\begin{aligned} &Gu_{l|k} \leq g \\ &\mathbb{P}\left\{[H]_j x_{l+1|k} \leq [h]_j \mid x_{l|k}\right\} \geq 1 - \varepsilon_j \end{aligned} \tag{5.18}$$

with $G \in \mathbb{R}^{q\times m}$, $g \in \mathbb{R}^q$ and $H \in \mathbb{R}^{p\times n}$, $h \in \mathbb{R}^p$. Finally, the stage cost ℓ is considered to be a quadratic function $\ell(x,u) = x^\top Qx + u^\top Ru$ with $Q \succ 0$, $R \succ 0$.

Similar to Section 3.1 the predicted state is divided into a deterministic, nominal part $z_{l|k} = \mathbb{E}_k\{x_{l|k}\}$ and a martingale $e_{l|k} = x_{l|k} - z_{l|k}$. The predicted inputs are parameterized by $u^p_{l|k} = Ke_{l|k} + v_{l|k}$, with decision variable $p_{l|k} = v_{l|k} \in \mathbb{R}^m$ and constant $K \in \mathbb{R}^{m\times n}$ such that $A_{cl} = A + BK$ is Schur stable. The parameterized stochastic MPC optimization (5.12) can then equivalently be written as

$$\tilde{V}_N(x_k) = \min_{\mathbf{p}_{N|k}} \sum_{l=0}^{N-1} \ell(z_{l|k}, v_{l|k}) + \ell_l \tag{5.19a}$$

$$\text{s.t. } z_{l+1|k} = Az_{l|k} + Bv_{l|k}, \qquad z_{0|k} = x_k \tag{5.19b}$$

$$z_{l|k} \in \mathbb{Z}_l \qquad l \in \mathbb{N}_1^N \tag{5.19c}$$

$$v_{l|k} \in \mathbb{V}_l \qquad l \in \mathbb{N}_0^{N-1} \tag{5.19d}$$

with constants $\ell_l = \mathbb{E}_k\{e_{l|k}^\top(Q + K^\top RK)e_{l|k}\}$ and constraint sets $\mathbb{Z}_l = \{z \mid Hz \leq \eta_l\}$, $\mathbb{V}_l = \{v \mid Gv \leq \mu_l\}$ where η_l, μ_l are defined according to Section 3.1.2 with $\varepsilon_{\bar{f}} = 0$.

Proposition 5.7 (Cost upper bound). *Let $\mathbb{Z}_\infty \subseteq \cap_{l=0}^\infty \mathbb{Z}_l$, $\mathbb{V}_\infty \subseteq \cap_{l=0}^\infty \mathbb{V}_l$, and $\mathbf{v}^*_{N|k}$, $\mathbf{z}^*_{N|k}$ be the optimal solution and predicted state trajectory of (5.19) with horizon length N and $\mathbb{Z}_l$, $\mathbb{V}_l$ replaced by $\mathbb{Z}_\infty$, $\mathbb{V}_\infty$, respectively. Assume $0 \in \mathbb{Z}_\infty$, $0 \in \mathbb{V}_\infty$. Assumption 5.4 is satisfied with $\ell_l^N = \ell_l$ and constants $\bar{a}, d \in \mathbb{R}_{\geq 0}$ if for all $N \in \mathbb{N}_{>0}$ and $e \in \mathbb{W}_\infty = \bigoplus_{l=0}^\infty A_{cl}^l B_w \mathbb{W}$ it holds*

$$\sum_{l=0}^{N-1} \ell(z^*_{l|k} + A_{cl}^l e, v^*_{l|k} + KA_{cl}^l e) \leq \bar{a}\sigma(z_{0|k} + e) + d.$$

Proof. For $x_k = 0$ and $N \in \mathbb{N}_{>0}$ the optimal solution is given by $v_{l|k} = 0$ and hence $\ell_l^N = \ell_l$. Let $\tilde{Q} = Q + K^\top RK$, then

$$\ell_l = \mathbb{E}\{\|e_{l|k}\|^2_{\tilde{Q}}\} = \mathbb{E}\{\|\sum_{i=0}^{l-1} A_{cl}^i B_w W_i\|^2_{\tilde{Q}}\} = \sum_{i=0}^{l-1} \mathbb{E}\{\|A_{cl}^i B_w W_i\|^2_{\tilde{Q}}\},$$

where the last equality follows from the independence assumption on the disturbance W_i. This implies satisfaction of $\ell_{N-l}^N \geq \ell_{N'-l}^{N'}$ for $N \geq N'$ as $\ell_i \geq \ell_j$ for $i \geq j$.

To derive $\bar{a}$ consider the feasible set of optimization (5.13) for linear systems and polytopic constraints, i.e.,

$$\mathbb{D}_{N,j}(x_{N-j|k}) = \left\{ v_{N-j|k}, \dots, v_{N-1|k} \;\middle|\; \begin{array}{l} \exists z_{N-j|k}, \dots, z_{N|k} \text{ satisfying} \\ z_{N-j|k} = \mathbb{E}_k\{x_{N-j|k}\} \\ \text{(5.19b)-(5.19d)} \; \forall l \in \mathbb{N}_{N-j}^{N-1} \end{array} \right\},$$

and the feasible set for (5.19) with $N = j$ and $\mathbb{Z}_l$, $\mathbb{V}_l$ replaced by $\mathbb{Z}_\infty$, $\mathbb{V}_\infty$, i.e.,

$$\mathbb{D}_j^\infty(x_k) = \left\{ v_{0|k}, \dots, v_{j-1|k} \;\middle|\; \begin{array}{l} \exists z_{0|k}, \dots, z_{j|k} \text{ satisfying} \\ \text{(5.19b)}, \; z_{0|k} = \mathbb{E}\{x_k\} \\ z_{l|k} \in \mathbb{Z}_\infty, \; v_{l|k} \in \mathbb{V}_\infty \; \forall l \in \mathbb{N}_0^{j-1} \end{array} \right\}.$$

Since $\mathbb{Z}_\infty \subseteq \mathbb{Z}_l$ and $\mathbb{V}_\infty \subseteq \mathbb{V}_l$ clearly $\mathbb{D}_j^\infty(x) \subseteq \mathbb{D}_{N,j}(x)$ and the solution $\mathbf{v}_{N|k}^*$ is feasible, but possibly suboptimal for the optimization (5.13).

For $l \geq 0$ the prediction error $e_{l+N-j|k} \sim A_{cl}^l e_{N-j|k} + e_{l|k}$ and therefore

$$\begin{aligned} &\mathbb{E}_k\{\ell(x_{l+N-j|k}, u_{l+N-j|k})\} \\ &= \mathbb{E}_k\{\ell(z_{l+N-j|k} + A_{cl}^l e_{N-j|k} + e_{l|k}, v_{l+N-j|k} + K(A_{cl}^l e_{N-j|k} + e_{l|k}))\} \\ &= \mathbb{E}_k\{\ell(z_{l+N-j|k} + A_{cl}^l e_{N-j|k}, v_{l+N-j|k} + KA_{cl}^l e_{N-j|k})\} + \ell_l. \end{aligned}$$

If for all $j \in \mathbb{N}$ and $e \in \mathbb{W}_\infty$ it holds

$$\sum_{l=0}^{j-1} \ell(z_{l|k}^* + A_{cl}^l e, v_{l|k}^* + KA_{cl}^l e) \leq \bar{a}\sigma(z_{0|k} + e) + d$$

it follows

$$\tilde{V}_{N,j}(x_{N-j|k}) \leq \bar{a}\mathbb{E}_k\{\sigma(x_{N-j|k})\} + \sum_{l=0}^{j-1} \ell_l^j + d$$

since $e_{N-j|k} \in \mathbb{W}_\infty$ $\mathbb{P}$ *a.s.*. ∎

Proposition 5.7 shows that for linear systems with the given input parameterization, finding an upper bound that satisfies Assumption 5.4 is of similar complexity as determining the respective upper bound for stabilizing nominal MPC without terminal constraint. Note that similar results can be derived for any input parameterization that is affine in the disturbance and the bound can be computed by drawing upon results on multi-parametric quadratic programming.

To conclude this section, we apply the result to the previously presented example.

Example 5.8 (Example 5.6 cont'd). Consider again the system in Example 5.6 with $W_k \sim \mathcal{U}[-0.3, 0.3]$, constraints given by

$$|u_{l|k}| \leq 1, \qquad \mathbb{P}\{x_{l+1|k} \leq 3 \mid x_{l|k}\} \geq 0.75, \qquad \mathbb{P}\{x_{l+1|k} \geq -3 \mid x_{l|k}\} \geq 0.75,$$

and prestabilizing control gain $K = K_{LQR} = 0.62$. The set $\mathbb{W}_\infty = [-0.49, 0.49]$ can be derived through a geometric series, leading to $\mathbb{Z}_\infty = \{z \in \mathbb{R} \mid |z| \leq 2.66\}$. Again, choosing $\sigma(x) = x^2$, an upper bound satisfying Assumption 5.4 is given by $\bar{a} = 1.59P$ and $d = 0.09$ which can be derived by optimization and noting that the optimal value function is a continuous, piecewise quadratic function. Note that, due to the constraints, we cannot choose $\bar{a} = P$, $d = 0$ as in Example 5.6 since $v_{l|k} = 0$ is not a feasible solution for all $z_{l|k} \in \mathbb{Z}_\infty$.

Stability with relaxed constraints

In stochastic MPC, one key to derive computationally tractable approximations as well as a significantly increased feasible region is the relaxation of the constraints (5.3) and (5.4) to

$$\begin{aligned} &\mathbb{P}(u_{l|k} \in \mathbb{U} \mid x_k) \geq 1 - \varepsilon_u, \\ &\mathbb{P}(x_{l+1|k} \in \mathbb{X} \mid x_k) \geq 1 - \varepsilon \end{aligned} \tag{5.20}$$

with $\varepsilon_u \in [0, 1)$. As shown in Chapter 3, this allows to approximate the stochastic optimal control problem by sampling approaches or general function approximation techniques used in uncertainty quantification, e.g., polynomial chaos expansion. Yet, since the chance constraint need not be satisfied for all future realizations x_{k+l} but only conditioned on the measured state x_k, at time $k+1$ this might lead to infeasibility of the optimal input trajectory computed at time k. As before, to prove stability, the probability of infeasibility of the candidate solution and possible cost increase needs to be taken into account explicitly.

In the following, consider the finite horizon optimal control program (5.5) but with relaxed constraints (5.20) and optimization over real-valued parameters $\mathbf{p}_{N|k} = (p_{l|k})_{l \in \mathbb{N}_0^{N-1}}$ instead of functions $\mathbf{u}_{N|k}$, i.e.,

$$\begin{aligned} \tilde{V}_N(x_k) = \min_{\mathbf{p}_{N|k}} \; & \mathbb{E}_k \left\{ \sum_{l=0}^{N-1} \ell(x_{l|k}, u_{l|k}(x_{l|k})) \right\} \\ \text{s.t.} \; & u_{l|k} = u^p_{l|k} \\ & (5.2), (5.20) \qquad l \in \mathbb{N}_0^{N-1}. \end{aligned} \tag{5.21}$$

Denote its minimizer by $\mathbf{p}^*_{N|k}$ and the set of admissible decision variables by

$$\mathbb{D}_N(x_k) = \left\{ \mathbf{p}_{N|k} \in \mathbb{R}^{m_p N} \;\middle|\; \begin{array}{l} \exists \mathbf{x}_{N|k} \text{ satisfying} \\ u_{l|k} = u^p_{l|k}, (5.2), (5.20) \; \forall l \in \mathbb{N}_0^{N-1} \end{array} \right\}.$$

Since $\mathbb{P}_k(x_{l|k} \in \mathbb{X}) \geq 1 - \varepsilon \not\Rightarrow \mathbb{P}_{k+1}(x_{l-1|k+1} \in \mathbb{X}) \geq 1 - \varepsilon$, equation (5.9) in the proof of Theorem 5.2 is invalidated as $u^{p^*}_{l|k}$ might not be an admissible input at time $k+1$. Define the candidate solution $\tilde{\mathbf{p}}_{N-1|k+1}$ as $\tilde{p}_{l|k+1} = p^*_{l+1|k}$ for all $l \in \mathbb{N}_0^{N-2}$ and let the probability of it not being admissible for the MPC optimization (5.21) with horizon length $N-1$ be given by

$$\epsilon_{\bar{f}}(x_k) = \mathbb{P}\left\{ \tilde{\mathbf{p}}_{N-1|k+1} \notin \mathbb{D}_{N-1}(f(x_k, u_k^{p^*}, W_k)) \right\}.$$

Furthermore, denote by $\mathbb{E}_k^{\bar{f}}\{A\} = \mathbb{E}_k\{A \mid \tilde{\mathbf{p}}_{N-1|k+1} \notin \mathbb{D}_{N-1}(f(x_k, u_k^{p^*}, W_k))\}$ the expected value of an event A given the candidate solution at time $k+1$ does not remain feasible.

Convergence of the closed loop can be proven if the maximal cost increase in the case of the candidate solution not being admissible as well as the probability of infeasibility $\epsilon_{\bar{f}}(x)$ can be bounded for all $x \in \mathbb{X}_\infty$.

Assumption 5.5 (Bounded cost increase). *There exists $\varepsilon_{\bar{f}} \in [0, 1)$ and $\delta : \mathbb{R}^n \to \mathbb{R}_{\geq 0}$ such that for all $x \in \mathbb{X}_\infty$ it holds $\varepsilon_{\bar{f}} \geq \epsilon_{\bar{f}}(x)$ and*

$$\mathbb{E}_k^{\bar{f}}\{J_N(x_{k+1})\} - J_N(x_k) \leq -\ell(x_k, u_k) + \delta(x_k).$$

In general, an upper bound δ satisfying Assumption 5.5 is given by $\delta(x) = \sup_{w \in \mathbb{W}, u \in \mathbb{U}} \bar{a}\sigma(f(x, u, w))$. Tighter bounds for linear systems with additive disturbances can be derived analogous to Chapter 3.1 through the Lipschitz constants of the optimal value function $\tilde{V}_N$ on the admissible set $\mathbb{X}_\infty$. For linear systems with multiplicative disturbance, an upper bound can be derived analogous to Chapter 3.2 by checking an LMI condition on the vertices of the admissible set. For general nonlinear systems with additive disturbances a bound given by a class $\mathcal{K}$ function in the diameter of the disturbance set $\mathbb{W}$ can be derived following the results of Grimm et al. (2007).

Under Assumption 5.5, qualitatively the same stability results as in Theorem 5.2 can be recovered for the approximate SMPC optimization with relaxed constraints. Under strengthened assumptions there exists a sufficiently large optimization horizon N such that the convergence results and the limiting behavior stated in Corollaries 2.5, 2.6, and 2.7 can be established.

Proposition 5.9 (Stability under relaxed constraints). *Consider the stochastic MPC algorithm based on optimization (5.21). Suppose Assumption 5.1, 5.2, 5.4, and 5.5 are satisfied and $x_0 \in \mathbb{X}_\infty$. For all $k \in \mathbb{N}$, the state and input trajectories of the closed-loop system (5.1) with SMPC feedback law $\kappa(x_k) = u^{p^*}_{0|k}(x_k)$ satisfy the input constraint $u_k \in \mathbb{U}$ almost surely and the state constraint $x_k \in \mathbb{X}$ with at least probability $1-\varepsilon$. There exists $c_1 \in \mathbb{R}_{\geq 0}$ such that*

$$\mathbb{E}_k\left\{\tilde{V}_N(x_{k+1})\right\} - \tilde{V}_N(x_k) \leq -\ell(x_k,u_k) + \varepsilon_{\bar{f}}\delta(x_k) + (1-\varepsilon_{\bar{f}})\frac{\bar{a}^2}{N-1}\sigma(x_k) + c_1.$$

If there exists $\lambda_l < 1$ with $\varepsilon_{\bar{f}}\delta(x_k) \leq \lambda_l\sigma(x_k) + d_2$, then there exists $N \in \mathbb{N}$ such that for all $x_0 \in \mathbb{X}_\infty$ it holds

$$\mathbb{E}_k\left\{\tilde{V}^0_N(x_{k+1})\right\} \leq \lambda\tilde{V}^0_N(x_k) + c_2$$

with $\lambda \in (0,1)$, $c_2 \in \mathbb{R}_{\geq 0}$, and $\tilde{V}^0_N(x) = \tilde{V}_N(x) - \sum_{l=0}^{N-1}\ell^N_l$.

Proof. The proof follows analogously to the proof of Theorem 5.2, using ideas presented in Chapter 3. Hence, we only highlight the main differences.

The input and state constraints (5.3), (5.4) are satisfied since only $u^*_{0|k}$ is applied which is deterministic and hence for all $k \in \mathbb{N}$ it holds $u_k \in \mathbb{U}$ and $\mathbb{P}_k\{x_{1|k} \in \mathbb{X}\} \geq 1-\varepsilon$. For the cost decrease, the argument follows by explicitly considering infeasibility of the candidate solution and using the law of total probability

$$\begin{aligned}
&\mathbb{E}_k\{V_N(x_{k+1})\} - V_N(x_k)\\
&= \epsilon_{\bar{f}}(x_k)\mathbb{E}^{\bar{f}}_k\{V_N(x_{k+1})\} + (1-\epsilon_{\bar{f}}(x_k))\mathbb{E}^f_k\{V_N(x_{k+1})\} - V_N(x_k)\\
&\leq \varepsilon_{\bar{f}}\mathbb{E}^{\bar{f}}_k\{V_N(x_{k+1}) - V_N(x_k)\} + (1-\varepsilon_{\bar{f}})\mathbb{E}^f_k\{V_N(x_{k+1}) - V_N(x_k)\}\\
&\leq -\ell(x_k,u_k) + \varepsilon_{\bar{f}}\delta(x_k) + (1-\varepsilon_{\bar{f}})\left(\frac{\bar{a}^2}{N-1}\sigma(x_k) + \bar{a}\bar{\ell} + d\right)
\end{aligned}$$

leading to $c_1 = (1-\varepsilon_{\bar{f}})(\bar{a}\bar{\ell} + d)$. For the second part note that if $N \geq \frac{(1-\varepsilon_{\bar{f}})\bar{a}^2}{1-\lambda_l} + 1$ then

$$\begin{aligned}
\mathbb{E}_k\{V^0_N(x_{k+1})\} - V^0_N(x_k) &\leq -\sigma(x_k) + \varepsilon_{\bar{f}}\delta(x_k) + (1-\varepsilon_{\bar{f}})\frac{\bar{a}^2}{N-1}\sigma(x_k) + c_1\\
&\leq \left(\frac{\lambda_l - 1}{\bar{a}} + (1-\varepsilon_{\bar{f}})\frac{\bar{a}}{N-1}\right)V^0_N(x_k) + c_2
\end{aligned}$$

which proves the statement with $\lambda = \frac{(1-\varepsilon_{\bar{f}})\bar{a}^2 + (\bar{a}+\lambda_l-1)(N-1)}{\bar{a}(N-1)}$ and $c_2 = c_1 + \varepsilon_{\bar{f}}d_2$. ∎

Note that similar to the results in Chapter 3, qualitatively the same asymptotic properties as for the conceptual algorithm can be recovered, but there are quantitative differences which can be observed especially when Corollary 2.5 and 2.6 are applied to Proposition 5.9. When using simplifications and relaxed constraints, the constants c_1 and c_2 are larger, i.e., the asymptotic average distance of the state to the origin increased. Similarly λ is smaller, which implies that the probability of leaving given sublevel sets is increased. Finally, the lower bound on the MPC horizon that is sufficient to prove convergence of the closed-loop is increased.

In this section, we have derived stability results for computationally tractable stochastic MPC algorithms that do not rely on terminal constraints and a terminal cost. In particular, we have discussed the usual simplifications of using an input parameterization and relaxing the state and input constraints, which where both considered in Chapter 3, as well. Unlike the transition from Chapter 2 to Chapter 3, the stability proof of the conceptual algorithm did not directly extend, but further assumptions were necessary.

5.4 Discussion and summary

In the following, we briefly discuss the limitations of the theoretical results and their implication on practical implementations. Finally, we conclude this chapter on stochastic MPC without terminal constraints and cost by summarizing the main results.

Limitations of the proposed approach

Similar to the stability results for nominal MPC without terminal constraints, the assumptions made in this chapter are in practice often difficult to verify rigorously. While the controllability Assumption 5.3 ii) reads similar to the nominal case, it is qualitatively different since a stochastic optimal control problem instead of a nominal, static optimization program is considered. For the practically relevant Assumption 5.4 ii) the question is further complicated by requiring the bound not only to hold for all $x \in \mathbb{X}_\infty$ but for all possible predicted state distributions $x_{l|k}$. However, a sufficient condition can be derived by checking the condition pointwise, which implies that it holds when evaluating the expected value. Furthermore, for linear systems we explicitly derived a sufficient condition, which is of similar complexity to checking the respective condition in the nominal case.

From an implementation point of view, rigorously checking these assumptions is necessary to derive a bound on the horizon length to guarantee convergence of the closed loop. This leads to the second limitation, which is of conceptual nature but also has severe implications on the practical implementation. The derived results are only sufficient but not necessary for convergence. In applications convergence is often observed with a much shorter horizon, which in turn implies that the derived bounds are generally conservative. This observation holds similarly for the nominal case and calls for further research in the field of stability without terminal constraints and cost.

Summary

The purpose of this chapter was to derive a stabilizing stochastic MPC algorithm without terminal constraints and without a terminal cost. Following the general theme of this thesis, we first presented a conceptual algorithm, which allowed to prove stability guarantees. Based upon these results, in the second part, we derived and discussed computationally tractable algorithms, which inherit similar convergences properties.

Analogous to Chapter 2.2, the conceptual algorithm was based upon an optimization over general feedback laws and an exact propagation of the probability density functions of the predicted states. The stability proof was then derived along the lines of the nominal case with the main difference being that of having a cost that is not necessarily positive definite with respect to a steady-state distribution as well as using conditional probabilities based upon the most recent measurement. While this allowed for local approximations of the online optimization as presented in Chapter 3, we highlighted that the stability proofs do not extend trivially to these computationally tractable relaxations as the principle of optimality cannot be applied.

In the second part, based upon these insights, the assumptions were strengthened to derive stability results for the simplified MPC algorithms. The basic principles were on the one hand to explicitly consider the "tail cost" of the MPC optimization to replace the principle of optimality and on the other hand, similar to Chapter 3, to consider the probability of infeasibility of the candidate solution and then to apply the law of total probability.

The general analysis carried out in this chapter yields rather conservative bounds for a sufficient stabilizing horizon length. However, the results give a good guideline on how to choose not only the horizon length, but the stage cost in the first place. To the best of our knowledge, the presented stability results for stochastic MPC algorithms without terminal ingredients are the first on this topic and lead to many open research questions, which are discussed in the following chapter.

Chapter 6

Conclusions

6.1 Discussion and summary

Model predictive control is by now an indispensable part of modern control theory and essential in the toolbox of any advanced control engineer. Unlike most conventional methods, it provides a constructive approach to controller design for multivariable systems while explicitly incorporating state and input constraints. However, disturbances and plant-model mismatch still provide a major challenge as they can significantly deteriorate the performance and even lead to instability if not taken into account explicitly. For this reason, there is a strong interest in MPC algorithms that explicitly include a disturbance or uncertainty model and significant progress has been made in recent years in this respect. Nonetheless, the topic has not yet been fully explored and has not reached a maturity, where it can reliably be applied with consistent results over a variety of different problem setups.

The results in this thesis contribute to the theoretical foundation of predictive control under uncertainty as well as to a practitioners toolbox for addressing constrained control problems. In particular, they are focused on systems with stochastic disturbance models or constant uncertainty. For the former, the results provide theoretically sound methods to guarantee constraint satisfaction up to a user-defined violation probability. For the latter, the proposed solutions guarantee robust constraint satisfaction while the model is learnt online to increase the performance.

The theoretical background was built in Chapter 2, where a conceptual stochastic model predictive control algorithm was presented and relevant closed-loop properties were proven. Equally important, Chapter 2 explicitly highlights the interplay of predictions modeled as random variables and realizations of the controlled process revealed by measurements. This difference is critical in the derivation of rigorous computational approaches presented in Chapter 3 and key to the discussion of recursive feasibility. Moreover, we briefly touched upon output feedback in the conceptual framework, which was then revisited in Chapter 4, where uncertainty in the model has been addressed.

As the conceptual algorithm is based upon optimization over general feedback laws and exact propagation of the probability density functions, it is not practically feasible and calls for computational approaches tailored to specific system classes. In Chapter 3 these were developed for linear systems subject to additive and multiplicative disturbances, respectively. Therein, on a conceptual level, a relaxation of the chance constraints has been discussed which led to a significantly increased feasible region and thorough discussion of the requirements for recursive feasibility and stability. In particular, we showed that an additional viability constraint on the applied input suffices if convergence is proven by explicitly considering infeasibility of the candidate solution and using conditional expectations. Thereby, the presented solution for linear systems with stochastic additive disturbances extends and unifies the previously obtained results of Kouvaritakis et al. (2010) and Korda et al. (2011). For the proposed algorithm, we not only proved an asymptotic bound on the closed-loop performance, but asymptotic stability with probability one of a minimal robust positively invariant set.

Regarding the computational aspects, a broadly applicable approach based on finite sample approximations to the stochastic program has been presented for each problem setup. On the one hand this allows for a data-driven MPC design. On the other hand, this reduces the required online optimization to a deterministic linearly constrained quadratic program, making the stochastic MPC algorithms amenable to fast systems and limited hardware, cf. (Mammarella et al., 2018b), without sacrificing theoretic guarantees. This sampling approximation pointed to the important question of a sufficient sample size to achieve a user defined confidence. Regarding this topic, we have provided insight into the difference between online and offline sampling approximations. For the latter, we have shown that the question of finding a lower bound on the sample complexity can be reduced to the general question of a sample approximation being a subset of a suitably chosen set defined by chance constraints. This has been answered in both cases using scenario theory and statistical learning theory, respectively. Finally, compared to online sampling approaches, the proposed design has the advantage that the viability constraint can be explicitly computed in order to guarantee robust recursive feasibility and thereby a well-defined control law.

With respect to model uncertainty, the stochastic framework has been sacrificed for a mathematically rigorous yet computationally tractable deterministic approach. In Chapter 4, an adaptive MPC algorithm has been presented, where, analogous to a Bayesian filter as discussed in Chapter 2, a membership set as well as a point estimate for the uncertain model parameters are updated recursively. The proposed MPC algorithm is applicable to linear systems with unknown but bounded parametric uncertainty and additive disturbance. Under this assumption, we have proven a finite ℓ_2 gain and robust constraint satisfaction for the closed loop. Again, the online optimization has been reduced to a linearly constrained quadratic program and was proven to be recursively feasible. The presented solution provides a general framework and can be easily extended and combined with further results in the MPC literature. As a specific example, we discussed parameter convergence under a persistence of excitation condition as well as tracking of constant reference signals.

The theoretical results and numerical examples also point to one major drawback and source of conservatism: The assumption of a suitable terminal controller and a terminal constraint. In Chapter 5, convergence of stochastic MPC schemes without these terminal ingredients have been investigated. In line with the previous chapters, we first discussed conceptual aspects, which eventually led to a deeper understanding of sufficient stability conditions and paved the way to extending the obtained stability results to computational approaches. Specifically, for parameterized feedback laws and the use of relaxed chance constraints, sufficient conditions for closed-loop stability have been derived.

In summary, the results presented in this thesis provide novel insights on the conceptual aspects of stochastic MPC as well as mathematically well founded computational approaches to solving constrained control problems under stochastic disturbances and uncertainty.

6.2 Outlook

The results obtained in this thesis naturally point to further interesting research topics, ranging from further development of the theory to providing a proof of concept and validation in a practically relevant application.

Starting from the computational approaches developed in Chapters 3 and 4, the next obvious step is the application of the proposed stochastic MPC algorithm in an actual application. Of particular interest are thereby whether and how the assumptions can be rigorously verified and the practical feasibility of the involved set computations in case of a large state and input space. Regarding the former, specifically relevant is the possible use of measured data to derive the sample approximations as well as the quantification of an outer bound on the disturbance. Regarding the latter, an analogous derivation using ellipsoids instead of polytopes or additional simplifications might be necessary to handle the computational complexity. First application results for systems with multiplicative and additive disturbance models have been obtained in two aerospace examples presented in (Mammarella et al., 2018a,b).

The cited implementation highlighted some weaknesses and pointed to a series of further interesting theoretical questions. For a small probability of allowed constraint violation ε, the number of samples became too large, creating a highly complex constraint set. Further research should not only focus on simplifying the resulting constraint sets, but rather on new results regarding sample approximations for sets defined by chance constraints. Since the bound on the sample complexity provided in this thesis is only sufficient, finding an improved lower bound or proving necessity remains an important open problem. Of similar relevance is research on algorithmic changes such that a lower sample complexity is required by design. Some first results in this direction have recently been obtained by Alamo et al. (2018), where a two step procedure has been proposed.

In the cited experimental setup, the samples were created numerically based upon assumptions on the disturbance. Although the observed results were promising, the proposed sampling approach may as well perform poorly in another application as in general the actual disturbance distribution is never precisely known. For deterministic approximations, there are well-known results on distributionally robust optimization, see e.g., (Goh and Sim, 2010). It would be of high practical relevance to establish similar results for sampling approaches, in particular to investigate the inherent robustness against distributional uncertainty as well as to develop distributionally robust sampling-based optimization algorithms. Similarly, the assumption of the disturbances being independent and identically distributed might often not be satisfied in practice. It would be worthwhile to investigate sampling approaches beyond the usual approach of coloring filters.

Focusing on algorithmic results, chance constraints provide a structured way of handling soft constraints and deal with disturbances that would otherwise lead to overly conservative results in the case of a robust set-membership approach. However, due to bounding the violation probability pointwise in time, the ability of dealing with very unlikely outliers that are of large magnitude is still limited. Only very little research has been done considering the probability over sample paths instead. To address this issue, in (Lorenzen et al., 2017c), results have been presented based upon probabilities conditioned on the initial state, i.e., using $\mathbb{P}_0$ instead of $\mathbb{P}_k$. Similarly, in (Hewing and Zeilinger, 2018) using probabilities conditioned on the state from a previous time step in case of infeasibility has been mathematically analysed and led to promising results in a simulation study. Most recently, Yan et al. (2018) introduced a stochastic MPC algorithm with chance constraints that are defined as a discounted sum of violation probabilities on an infinite horizon. Considering this brief discussion, identifying a suitable constraint formulation and subsequently developing a theoretically sound as well as computationally tractable MPC algorithm appears to be a promising research direction of practical relevance.

In Chapter 4, we simplified the stochastic setup to a set-membership assumption.

Based upon the obtained results and insights, the next obvious step is to develop similar results in a stochastic framework. Further research could focus on either an adaptive MPC formulation with or without an explicit persistent excitation or on an objective function which naturally leads to "dual control" and thus an excitation only if it decreases the overall closed-loop cost. In both cases, even the distillation of a relevant and tractable problem setup provides a valuable contribution, which would probably attract significant attention and lead to a subsequent development of applicable algorithms.

The research on stochastic MPC without a terminal constraint and terminal cost was motivated by its easier design and an often increased admissible region. However, the assumptions made in Chapter 5 are in general difficult to verify rigorously and the obtained lower bound on the horizon length that is sufficient for convergence is in most cases rather large, which increases the computational complexity. Further research beyond current approaches is needed to fully understand necessary and sufficient stability conditions. Interestingly, most results for unconstrained nominal MPC explicitly rely on an optimal solution, an assumption that cannot be relied on for stochastic MPC. New paths should be explored not only for nominal models but also explicitly for systems under uncertainty and disturbances.

In summary, the results presented in this thesis do not only provide theoretical insights and algorithmic solutions for stochastic MPC, but also lead to many interesting new questions and open up different research directions, ranging from conceptual aspects to application of the computational approaches.

Appendix A

Stability of deterministic and stochastic system

In the following, we summarize the basic invariance and stability definitions that are used throughout this thesis. The definitions can be found in standard textbooks, see e.g., (Rawlings et al., 2017) for stability results tailored to the context of discrete-time nominal and robust MPC or (Kushner, 1967; Meyn and Tweedie, 2009; Doob, 1990) for stochastic systems and convergence of stochastic processes.

Nominal systems

Consider a deterministic, discrete-time system of the form

$$x_{k+1} = f(x_k) \tag{A.1}$$

with state $x_k \in \mathbb{R}^n$, given initial condition $x_0 \in \mathbb{R}^n$, and locally bounded transition function $f : \mathbb{R}^n \to \mathbb{R}^n$.

A point $\bar{x} \in \mathbb{R}^n$ is said to be an *equilibrium point* of the system dynamics (A.1) if $\bar{x} = f(\bar{x})$. A closed set $\mathbb{S}$ is said to be *positively invariant* or *forward invariant* under the system dynamics (A.1) if $f(x) \in \mathbb{S}$ for all $x \in \mathbb{S}$.

In the following definitions, we assume without loss of generality, that $\bar{x} = 0$ is an equilibrium point of the system. However, they directly generalize to arbitrary $\bar{x} \in \mathbb{R}^n$ and closed, positive invariant sets.

Definition A.1 (Asymptotic stability). *The origin is (locally)* stable *for* (A.1) *if, for each $\varepsilon \in \mathbb{R}_{>0}$, there exists $\delta \in \mathbb{R}_{>0}$ such that $\|x_0\| < \delta$ implies $\|x_k\| < \varepsilon$ for all $k \in \mathbb{N}$. It is* asymptotically stable *if it is stable and* $\lim_{k\to\infty} \|x_k\| = 0$ *for all x_0 in a neighborhood of the origin.*

The origin is said to be globally asymptotically stable if in the above definition convergence is satisfied for all $x_0 \in \mathbb{R}^n$. In the context of MPC, it is useful to restrict the definition from global convergence to convergence within a constrained set: Suppose $\mathbb{S}$ is positively invariant under (A.1) and $0 \in \mathbb{S}$. Then the origin is *stable in* $\mathbb{S}$ if Definition A.1 applies with the additional restriction of $x_0 \in \mathbb{S}$ and convergence for all $x_0 \in \mathbb{S}$. The set $\mathbb{S}$ is called *region of attraction*.

Asymptotic stability is generally proved by using a *Lyapunov function*. In particular, to prove Theorem 2.3 the optimal value function is employed as such.

Definition A.2 (Lyapunov function). *Suppose the set $\mathbb{S}$ is positive invariant and $0 \in \mathbb{S}$. A function $V : \mathbb{S} \to \mathbb{R}_{\geq 0}$ is a* Lyapunov function *in $\mathbb{S}$ for the system* (A.1) *if there exists functions $\alpha_1, \alpha_2 \in \mathcal{K}_\infty$ and a continuous, positive definite function $\alpha_3 : \mathbb{R} \to \mathbb{R}_{\geq 0}$ such that for any $x \in \mathbb{S}$*

$$\alpha_1(\|x\|) \leq V(x) \leq \alpha_2(\|x\|)$$
$$V(f(x)) - V(x) \leq -\alpha_3(\|x\|).$$

Theorem A.3 (Lyapunov function implies asymptotic stability). *Suppose $\mathbb{S}$ is positive invariant and $0 \in \mathbb{S}$. If V is a Lyapunov function in $\mathbb{S}$ for the system* (A.1), *then the origin is asymptotically stable in $\mathbb{S}$.*

Disturbed systems

In the following, the autonomous system (A.1) is extended with an exogenous input, that is we consider a discrete-time system of the form

$$x_{k+1} = f(x_k, w_k) \tag{A.2}$$

with external input $w_k \in \mathbb{R}^{m_w}$, state $x_k \in \mathbb{R}^n$, given initial condition $x_0 \in \mathbb{R}^n$, and continuous transition function $f : \mathbb{R}^n \times \mathbb{R}^{m_w} \to \mathbb{R}^n$.

Suppose a set $\mathbb{W} \subset \mathbb{R}^{m_w}$ is given such that $w_k \in \mathbb{W}$ for all $k \in \mathbb{N}$. A closed set $\mathbb{S}$ is said to be *robust positively invariant* or *robust forward invariant* under the system dynamics (A.1) if $f(x, w) \in \mathbb{S}$ for all $x \in \mathbb{S}$ and $w \in \mathbb{W}$.

In the following, $(w_k)_{k \in \mathbb{N}}$ is assumed to be a Markov chain. Analogous to the different types of convergence of random variables, different stability definitions are commonly used in the control literature and within this thesis to characterize an equilibrium point of (A.2). For a discussion of implications among the following and further stability concepts, the interested reader is referred to, e.g., (Kozin, 1969).

Definition A.4 (Asymptotic stability in probability). *The origin is said to be* stable in probability *for the system dynamics* (A.2) *if, for each $\varepsilon \in \mathbb{R}_{>0}$ and $\rho \in \mathbb{R}_{>0}$, there exists $\delta \in \mathbb{R}_{>0}$ such that $\|x_0\| < \delta$ implies*

$$\mathbb{P}\{\sup_{k \in \mathbb{N}} \|x_k\| > \varepsilon\} < \rho.$$

It is asymptotically stable in probability *if additionally for any* $\varepsilon_2 \in \mathbb{R}_{>0}$

$$\lim_{k' \to \infty} \mathbb{P}\{\sup_{k>k'} \|x_k\| > \varepsilon_2\} = 0$$

for all x_0 *in a neighborhood of the origin.*

Definition A.5 (Asymptotic stability in the mth mean). *The origin is said to be* stable in the mth mean *for the system dynamics* (A.2) *if the mth moments of* x_k *exist for all k and, for each* $\varepsilon \in \mathbb{R}_{>0}$*, there exists* $\delta \in \mathbb{R}_{>0}$ *such that* $\|x_0\| \le \delta$ *implies*

$$\mathbb{E}\{\sup_{k\ge 0} \|x_k\|_m^m\} < \varepsilon.$$

It is asymptotically stable in the mth mean *if additionally*

$$\lim_{k' \to \infty} \mathbb{E}\{\sup_{k\ge k'} \|x_k\|_m^m\} = 0.$$

for all x_0 *in a neighborhood of the origin.*

The case $m = 2$ has received most attention in the control literature and is generally called *mean square stability.*

Definition A.6 (Asymptotic stability with probability 1). *The origin is said to be* stable with probability 1 *for the system dynamics* (A.2) *if, for each* $\varepsilon \in \mathbb{R}_{>0}$

$$\lim_{\delta \to 0} \mathbb{P}_0\{\sup_{\|x_0\|<\delta} \sup_{k\in\mathbb{N}} \|x_k\| > \varepsilon\} = 0.$$

It is asymptotically stable with probability 1 *if additionally for any* $\varepsilon_2 \in \mathbb{R}_{>0}$,

$$\lim_{k' \to \infty} \mathbb{P}\{\sup_{k\ge k'} \|x_k\| > \varepsilon_2\} = 0$$

for all x_0 *in a neighborhood of the origin.*

Stability with probability 1 is often also called *almost sure stability.*

Similar to the deterministic case, the definitions extend to global asymptotic stability and constrained versions. An extension of Theorem A.3 to stochastic systems can be found in, e.g., Kushner (1967). However, it is worth highlighting that the above given stability definitions require that the effect of the disturbance goes to zero for the state going to the origin. As this is often not the case, asymptotic bounds on the moments of the state or explicit bounds on the probability that sample paths leave a region of interest are generally of equal if not superior interest.

Appendix B

Technical proofs

B.1 SMPC closed-loop properties

As to be expected, the proof of Proposition 2.4 follows the standard results given in Theorem 2.3 and can be found in, e.g., (Grüne and Pannek, 2017).

Proof. Let $\mathbf{u}_{N|k}$ be an admissible solution of (2.8) and consider the candidate solution $\tilde{\mathbf{u}}_{N|k+1}$ defined by $\tilde{u}_{l|k+1} \equiv u_{l+1|k}$ for $l \in \mathbb{N}_0^{N-2}$ and $\tilde{u}_{N-1|k+1} \equiv \kappa_f$. Furthermore, denote the corresponding state sequence resulting from (2.5) with the initial condition x_{k+1} by $\tilde{\mathbf{x}}_{N|k+1}$.

Hence, the candidate solution $\tilde{\mathbf{u}}_{N|k+1}$ and state sequence $\tilde{\mathbf{x}}_{N|k+1}$ satisfy the dynamic constraint (2.8b). Furthermore, for $l \in \mathbb{N}_0^{N-1}$ the distribution of $\tilde{x}_{l|k+1}$ is equivalent to the conditional distribution of $x^*_{l+1|k}$ given the disturbance realization w_k. This implies $\mathbb{P}_w$ *a.s.* satisfaction of the state constraints (2.8c) by the candidate solution for $l \in \mathbb{N}_0^{N-1}$ since

$$\begin{aligned}
&\mathbb{P}\{x^*_{l+1|k} \in \mathbb{X} \mid x^*_{l|k}\} \geq 1-\varepsilon \quad \mathbb{P}_k \; a.s.\\
\overset{\mathbb{P}_w\ a.s.}{\Longrightarrow}\ &\mathbb{P}\{x^*_{l+1|k} \in \mathbb{X} \mid x^*_{l|k}, w_k\} \geq 1-\varepsilon \quad \mathbb{P}_k \; a.s.\\
\Longleftrightarrow\ &\mathbb{P}\{\tilde{x}_{l|k+1} \in \mathbb{X} \mid \tilde{x}_{l-1|k+1}\} \geq 1-\varepsilon \quad \mathbb{P}_{k+1} \; a.s.
\end{aligned}$$

and similarly $\tilde{x}_{N-1|k+1} \in \mathbb{X}_f$ $\mathbb{P}_{k+1}$ *a.s.* as well as satisfactions of the input constraint (2.8d) for $l \in \mathbb{N}_0^{N-2}$. The latter in turn implies by Assumption 2.6 satisfaction of the terminal constraint (2.8e), satisfaction of the state constraint for $l = N$, and satisfaction of the input constraint for $l = N-1$. In summary, this proves forward invariance of the admissible set.

Satisfaction of the state and input constraints follows trivially from feasibility of the SMPC optimization (2.8), in particular the optimal solution satisfying (2.8c) and (2.8d) for $l = 0$.

To prove (2.9) note that

$$
\begin{aligned}
&\mathbb{E}_k \left\{V_N(x_{k+1})\right\} - V_N(x_k) \\
&= \mathbb{E}_k \left\{ \mathbb{E}_{k+1} \left\{ \sum_{l=0}^{N-1} \ell(x^*_{l|k+1}, u^*_{l|k+1}(x^*_{l|k+1})) + V_f(x^*_{N|k+1}) \right\}\right\} - V_N(x_k) \\
&\leq \mathbb{E}_k \left\{ \mathbb{E}_{k+1} \left\{ \sum_{l=0}^{N-1} \ell(\tilde{x}_{l|k+1}, \tilde{u}_{l|k+1}(\tilde{x}_{l|k+1})) + V_f(\tilde{x}_{N|k+1}) \right\}\right\} - V_N(x_k) \\
&\leq \mathbb{E}_k \left\{ \mathbb{E}_{k+1} \left\{ \sum_{l=0}^{N-2} \ell(\tilde{x}_{l|k+1}, \tilde{u}_{l|k+1}(\tilde{x}_{l|k+1})) + V_f(\tilde{x}_{N-1|k+1}) + c_1 \right\}\right\} - V_N(x_k) \\
&= \mathbb{E}_k \left\{ \sum_{l=1}^{N-1} \ell(x^*_{l|k}, u^*_{l|k}(x^*_{l|k})) + V_f(x^*_{N|k}) \right\} - V_N(x_k) + c_1 \\
&= -\ell(x_k, u_k) + c_1.
\end{aligned}
$$

The first inequality follows from possible suboptimality of the candidate solution, the second inequality follows from Assumption 2.6 iii), and the second equality follows from taking iterated expectations.

Finally, note that $\mathbb{E}_k \left\{V_N(x_{k+1})\right\} - V_N(x_k) = \mathbb{E}_k \left\{V^0_N(x_{k+1})\right\} - V^0_N(x_k)$ and hence

$$
\begin{aligned}
\mathbb{E}_k \left\{ V^0_N(x_{k+1}) \right\} &\leq V^0_N(x_k) - \ell(x_k, u_k) + c_1 \\
&\leq V^0_N(x_k) - \sigma(x_k) + c_1 \\
&\leq (1 - \frac{1}{\bar{a}}) V^0_N(x_k) + c_1,
\end{aligned}
$$

which proves (2.10) with $\lambda = (1 - \frac{1}{\bar{a}})$. ■

B.2 VC-Dimension of the class of linear half-spaces

In the discussed application, the origin is always within the constraint set. Therefore we explicitly consider half-spaces that include the origin $\{x \in \mathbb{R}^n \mid ax \leq 1\}$, which reduces the VC-Dimension by 1 compared to arbitrary half-spaces.

Proposition B.1. *The VC-Dimension of the class of linear half-spaces that include the origin* $\mathbb{H}_0 = \{x \in \mathbb{R}^d \mid ax \leq 1\}$, $a \in \mathbb{R}^{1\times d}$ *is less or equal than* d.

Proof. We prove the Proposition by showing that a set $\mathbb{X} = \{x_1, \dots, x_{d+1}\}$ of cardinality $d+1$ cannot be shattered by $\mathbb{H}_0$. Let $\mathbb{X}$ be given. By Radon's Theorem, the set $\tilde{\mathbb{X}} = \{0\} \cup \mathbb{X}$ can be partitioned into two sets $\mathbb{X}_1$ and $\mathbb{X}_2$ such that $\mathrm{co}(\mathbb{X}_1) \cap$

$\text{co}(\mathbb{X}_2) \neq \emptyset$, where $\text{co}(\mathbb{X})$ denotes the convex hull of the elements of $\mathbb{X}$. Without loss of generality assume $0 \in \mathbb{X}_1$. To derive a contradiction, assume there exists $a \in \mathbb{R}^{1\times d}$ such that $ax_i \leq 1$ for all $x_i \in \mathbb{X}_1$ and $ax_j > 1$ for all $x_j \in \mathbb{X}_2$, hence $ax \leq 1$ for all $x \in \text{co}(\mathbb{X}_1)$ and $ax > 1$ for all $x \in \text{co}(\mathbb{X}_2)$ which contradicts $\text{co}(\mathbb{X}_1) \cap \text{co}(\mathbb{X}_2) \neq \emptyset$. ∎

B.3 Persistence of excitation

Proof of Lemma 4.19. Let x_k be the solution of (4.1) with $w_k \equiv 0$. By (Green and Moore, 1986, Corollary 2.4) $([u_k^\top \ \cdots \ u_{k-n}^\top])_{k\geq n}$ being PE implies $([x_k^\top, u_k^\top])_{k\geq n}$ being PE. Let $d_k^i = [D_k]_i$,

$$\Xi_i = \begin{bmatrix} [A_1 \quad B_1]_i \\ \cdots \\ [A_p \quad B_p]_i \end{bmatrix}, \text{ and } S_{k_0} = \sum_{k=k_0}^{k_0+P-1} \begin{bmatrix} x_k \\ u_k \end{bmatrix} \begin{bmatrix} x_k^\top & u_k^\top \end{bmatrix}.$$

Then

$$\begin{aligned} \sum_{k=k_0}^{k_0+P-1} D_k^\top D_k &= \sum_{k=k_0}^{k_0+P-1} \sum_{i=1}^{n} {d_k^i}^\top d_k^i \\ &= \sum_{k=k_0}^{k_0+P-1} \sum_{i=1}^{n} \Xi_i \begin{bmatrix} x_k \\ u_k \end{bmatrix} \begin{bmatrix} x_k^\top & u_k^\top \end{bmatrix} \Xi_i \\ &= \sum_{i=1}^{n} \Xi_i S_{k_0} \Xi_i. \end{aligned}$$

Let $\underline{\sigma}$ and $\bar{\sigma}$ be the smallest and largest singular value of

$$\begin{bmatrix} [A_1 \quad B_1]_1 & \ldots & [A_1 \quad B_1]_n \\ & \cdots & \\ [A_p \quad B_p]_1 & \ldots & [A_1 \quad B_1]_n \end{bmatrix},$$

which are nonzero by Assumption 4.4, and define $\alpha' = \underline{\sigma}^2\alpha$ and $\beta' = \bar{\sigma}^2\beta$. Given $\alpha I \preceq S_{k_0} \preceq \beta I$, we have $\alpha' I \preceq \sum_{i=1}^{n} \Xi_i S_{k_0} \Xi_i \preceq \beta' I$ which proves the claim. ∎

Appendix C

Miscellaneous

C.1 Explicit solution matrices

For completeness, the explicit solution and cost matrices used in Section 3.2 are given in the following. We use I_n to denote the $n \times n$ identity matrix and $0_{n\times m}$ to denote a zero matrix of dimension $n \times m$.

Solution matrices The matrices $\Phi^0_{l|k}$, $\Phi^u_{l|k}$, and Γ_l are obtained by solving the dynamics (3.23) with prestabilizing input (3.25) explicitly for the predicted state $x_{l|k}$ and input $u_{l|k}$. For $l = 0$ we obtain $\Phi^0_{0|k} = I_n$ and $\Phi^u_{0|k} = 0_{n\times mN}$. For $l \geq 1$ with the notation $A^{cl}_{l|k} = A(W_{l+k}) + B(W_{l+k})K$ we have

$$\Phi^0_{l|k} = A^{cl}_{l-1|k} A^{cl}_{l-2|k} \cdots A^{cl}_{0|k},$$
$$\Phi^u_{l|k} = \begin{bmatrix} A^{cl}_{l-1|k} \cdots A^{cl}_{1|k} B_{0|k} & \dots & B_{l-1|k} & 0_{n\times(N-l)m} \end{bmatrix}.$$

The matrix $\Gamma_l = \begin{bmatrix} 0_{m\times lm} & I_m & 0_{m\times(N-l-1)m} \end{bmatrix}$ selects the l-th entry in the stack vector $\mathbf{v}_{N|k}$.

Cost matrix Let

$$\Phi_N(\mathbf{W}_{N|k}) = \begin{bmatrix} \Phi^0_{0|k} & \Phi^u_{0|k} \\ \vdots & \vdots \\ \Phi^0_{N|k} & \Phi^u_{N|k} \end{bmatrix}, \quad \Gamma = \begin{bmatrix} 0_{mN\times n} & I_{mN} \end{bmatrix},$$

$\bar{Q} = I_N \otimes Q$, $\bar{R} = I_N \otimes R$ and $\bar{K} = I_N \otimes K$. The explicit cost matrix $\tilde{Q}$ in (3.40) is then given by

$$\begin{aligned} \tilde{Q} = & \mathbb{E}\Big\{ \Phi_N(\mathbf{W}_{N|k})^\top \begin{bmatrix} \bar{Q} & 0_{nN\times n} \\ 0_{n\times nN} & P \end{bmatrix} \Phi_N(\mathbf{W}_{N|k}) \\ & + \left[\bar{K}\Phi_{N-1}(\mathbf{W}_{N|k}) + \Gamma\right]^\top \bar{R} \left[\bar{K}\Phi_{N-1}(\mathbf{W}_{N|k}) + \Gamma\right] \Big\}, \end{aligned} \tag{C.1}$$

where the expected value can be solved to the desired accuracy by using a suitable numerical integration rule.

Notation

General mathematical notation

$*$	convolution operator
$\sim$	distributed according to, e.g., $W \sim \mathbb{P}_w$
$\otimes$	Kronecker product
$\oplus$	Minkowsky sum
$\ominus$	Pontryagin difference
$<, \leq, >, \geq$	element-wise inequalities for vectors and matrices
$\succ, \succeq, \prec, \preceq$	relation operator for real symmetric matrices, $A \succ B$ if $A - B$ positive definite
$[x]_i$	i-th entry of a vector $x \in \mathbb{R}^n$
$[H]_i$	i-th row of a matrix $H \in \mathbb{R}^{n \times m}$
$\|x\|_Q$	$\|x\|_Q := \sqrt{x^\top Q x}$, weighted 2-norm for given matrix $Q \succ 0$
$\|x\|_y$	$\|x\|_y := d(x, y)$, distance between two points x and y in a metric space with metric d
$\|x\|_{\mathbb{A}}$	$\|x\|_{\mathbb{A}} := \min_{y \in \mathbb{A}} d(x, y)$, distance between a point x and a closed set $\mathbb{A}$ in a metric space with metric d
$0_{n \times m}$	$n \times m$ matrix of zeros
$\mathbb{1}_{\mathbb{A}}(\cdot)$	indicator function on the set $\mathbb{A}$
$\mathcal{B}$	Borel sigma algebra
$\mathbb{B}_\varepsilon$	open ball of radius ε in an Euclidean space
$\mathrm{co}\{x_1, \ldots, x_n\}$	convex hull of points $x_1, \ldots, x_n$
δ	Dirac delta function
$\mathrm{diag}(a, b, c, \ldots)$	diagonal matrix with entries a, b, c, ... on the diagonal
e	Euler's number
$\mathbb{E}_k\{\cdot\}$	$\mathbb{E}_k\{\cdot\} := \mathbb{E}\{\cdot \mid x_k\}$, conditional expectation given x_k
I_n	$n \times n$ identity matrix
$\mathcal{K}$	class of continuous, strictly increasing functions $\alpha : \mathbb{R}_{\geq 0} \to \mathbb{R}_{\geq 0}$ with $\alpha(0) = 0$
$\mathcal{K}_\infty$	class of functions $\alpha \in \mathcal{K}$ that are unbounded
$\mathcal{KL}$	class of continuous functions $\beta : \mathbb{R}_{\geq 0} \times \mathbb{N}$ with $\beta(\cdot, k) \in \mathcal{K}$ for any $k \in \mathbb{N}$ and $\beta(x, \cdot)$ decreasing with $\lim_{k \to \infty} \beta(x, k) = 0$ for any $x \in \mathbb{R}_{\geq 0}$
$\lambda_{\min}(A)$, $\lambda_{\max}(A)$	minimal and maximal eigenvalue of a matrix A
$\mathbb{N}$, $\mathbb{N}_{>0}$	natural numbers including and excluding zero
$\mathbb{N}_a^b$	natural numbers in the interval $[a, b]$
$\Pi_{\mathbb{A}}(x)$	Euclidean projection of a vector x onto a closed set $\mathbb{A}$
$\mathbb{P}_k\{\cdot\}$	$\mathbb{P}_k\{\cdot\} := \mathbb{P}\{\cdot \mid x_k\}$, conditional probability given x_k
$\mathbb{R}$, $\mathbb{R}_{\geq 0}$, $\mathbb{R}_{>0}$	real, non-negative real, and strictly positive real numbers
$\mathbb{R}^n$	real-valued n-vectors
$\mathbb{R}^{n \times m}$	real-valued $n \times m$ matrices

MPC variables and symbols

J_N	MPC cost function
κ	MPC control law
κ_f	terminal control law
ℓ	running cost
m	input dimension
n	state dimension
N	MPC horizon length
$u_{l\|k}$	input prediction at time k for time $l+k$
$\mathbf{u}_{N\|k}$	$\mathbf{u}_{N\|k} = (u_{l\|k})_{l\in\mathbb{N}_0^{N-1}}$, sequence of N predicted inputs, predicted at time k
$\mathcal{U}$	input space
$\mathbb{U}$	input constraint set
V_f	terminal cost
V_N	MPC optimal value function
$x_{l\|k}$	state prediction at time k for time $l+k$
$\mathbf{x}_{N\|k}$	$\mathbf{x}_{N\|k} = (x_{l\|k})_{l\in\mathbb{N}_1^{N}}$, sequence of N predicted states, predicted at time k
$\mathcal{X}$	state space
$\mathbb{X}$	state constraint set
$\mathbb{X}_f$	terminal constraint set
$\mathbb{X}_N$	feasible set for the MPC controller with horizon N
$\mathbb{Z}$	mixed state and input constraint set

Acronyms and abbreviations

iid	independent and identically distributed
LQR	linear quadratic regulator
LP	linear program
$\mathbb{P}$ *a.s.*	almost surely with respect to distribution $\mathbb{P}$
pdf	probability density function
QP	quadratic program
(S)MPC	(stochastic) model predictive control

Bibliography

V. Adetola and M. Guay. Robust adaptive MPC for constrained uncertain nonlinear systems. *International Journal of Adaptive Control and Signal Processing*, 25(2): 155–167, 2011.

M. Alamir and G. Bornard. Stability of a truncated infinite constrained receding horizon scheme: the general discrete nonlinear case. *Automatica*, 31(9):1353–1356, 1995.

T. Alamo, R. Tempo, D. R. Ramírez, and E. F. Camacho. A new vertex result for robustness problems with interval matrix uncertainty. *Systems & Control Letters*, 57(6):474–481, 2008.

T. Alamo, R. Tempo, and E. F. Camacho. Randomized strategies for probabilistic solutions of uncertain feasibility and optimization problems. *IEEE Transactions on Automatic Control*, 54(11):2545–2559, 2009.

T. Alamo, V. Mirasierra, F. Dabbene, and M. Lorenzen. Safe approximations of chance constraints by partial immersion and probabilistic scaling. In *Proceedings of the IEEE Conference on Decision and Control*, 2018. submitted.

R. Andersson. Adaptive and dual model predictive control. Master's thesis, University of Stuttgart, Institute for Systems Theory and Automatic Control, 2018. unpublished.

A. Aswani, H. Gonzalez, S. S. Sastry, and C. Tomlin. Provably safe and robust learning-based model predictive control. *Automatica*, 49(5):1216–1226, 2013.

N. Athanasopoulos, G. Bitsoris, and M. Lazar. Construction of invariant polytopic sets with specified complexity. *International Journal of Control*, 87(8):1681–1693, 2014.

J.-P. Aubin, P. Saint-Pierre, and A. M. Bayen. *Viability theory: new directions*. Springer, Berlin, Germany, 2nd edition, 2011.

E.-W. Bai, H. Cho, and R. Tempo. Convergence properties of the membership set. *Automatica*, 34(10):1245–1249, 1998.

I. Batina. *Model predictive control for stochastic systems by randomized algorithms*. PhD thesis, TU Eindhoven, 2004.

D. S. Bayard and A. Schumitzky. Implicit dual control based on particle filtering and forward dynamic programming. *International Journal of Adaptive Control and Signal Processing*, 24(3):155–177, 2010.

R. Bellman. *Dynamic Programming*. Princeton University Press, 1957.

A. Bemporad, M. Morari, V. Dua, and E. N. Pistikopoulos. The explicit linear quadratic regulator for constrained systems. *Automatica*, 38(1):3–20, 2002.

D. Bernardini and A. Bemporad. Stabilizing model predictive control of stochastic constrained linear systems. *IEEE Transactions on Automatic Control*, 57(6):1468–1480, 2012.

D. P. Bertsekas. Infinite time reachability of state-space regions by using feedback control. *IEEE Transactions on Automatic Control*, 17(5):604–613, 1972.

D. P. Bertsekas and S. E. Shreve. *Stochastic Optimal Control: The Discrete Time Case*, volume 139 of *Mathematics in Science and Engineering*. Academic Press, New York, USA, 1978.

N. P. Bhatia and G. P. Szegő. *Dynamical Systems: Stability Theory and Applications*, volume 35. Springer Berlin Heidelberg, 1967.

R. R. Bitmead, M. Gevers, and V. Wertz. *Adaptive Optimal Control, The Thinking Man's GPC*. Prentice Hall, 1990.

L. Blackmore, M. Ono, A. Bektassov, and B. Williams. A probabilistic particle-control approximation of chance-constrained stochastic predictive control. *IEEE Transactions on Robotics*, 26(3):502–517, 2010.

F. Blanchini. Set invariance in control. *Automatica*, 35(11):1747–1767, 1999.

F. Blanchini and S. Miani. *Set-Theoretic Methods in Control*. Systems & Control: Foundations & Applications. Birkhäuser Boston, 2nd edition, 2015.

S. Boyd, L. E. Ghaoui, E. Feron, and V. Balakrishnan. *Linear Matrix Inequalities in System and Control Theory*, volume 15. Siam, 1994.

M. Bujarbaruah, X. Zhang, and F. Borrelli. Adaptive MPC with chance constraints for FIR systems. *ArXiv e-prints*, abs/1804.09790, 2018.

G. Calafiore. Random convex programs. *SIAM Journal on Optimization*, 20(6): 3427–3464, 2010.

G. Calafiore and M. C. Campi. The scenario approach to robust control design. *IEEE Transactions on Automatic Control*, 51(5):742–753, 2006.

G. Calafiore and L. Fagiano. Stochastic model predictive control of LPV systems via scenario optimization. *Automatica*, 49(6):1861–1866, 2013a.

G. Calafiore and L. Fagiano. Robust model predictive control via scenario optimization. *IEEE Transactions on Automatic Control*, 58(1):219–224, 2013b.

G. Calafiore, F. Dabbene, and R. Tempo. Research on probabilistic methods for control system design. *Automatica*, 47(7):1279–1293, 2011.

E. F. Camacho and C. B. Alba. *Model Predictive Control*. Springer London, 2007.

M. C. Campi and S. Garatti. The exact feasibility of randomized solutions of uncertain convex programs. *SIAM Journal on Optimization*, 19(3):1211–1230, 2008.

M. C. Campi and S. Garatti. A sampling-and-discarding approach to chance-constrained optimization: Feasibility and optimality. *Journal of Optimization Theory and Applications*, 148(2):257–280, 2011.

M. Cannon, B. Kouvaritakis, S. V. Raković, and Q. Cheng. Stochastic tubes in model predictive control with probabilistic constraints. *IEEE Transactions on Automatic Control*, 56(1):194–200, 2011.

M. Cannon, Q. Cheng, B. Kouvaritakis, and S. V. Raković. Stochastic tube MPC with state estimation. *Automatica*, 48(3):536–541, 2012.

H. Chen and F. Allgöwer. A quasi-infinite horizon nonlinear model predictive control scheme with guaranteed stability. *Automatica*, 34(10):1205–1217, 1998.

L. Chisci, A. Garulli, A. Vicino, and G. Zappa. Block recursive parallelotopic bounding in set membership identification. *Automatica*, 34(1):15–22, 1998.

L. Chisci, J. Rossiter, and G. Zappa. Systems with persistent disturbances: predictive control with restricted constraints. *Automatica*, 37(7):1019–1028, 2001.

D. Clarke, C. Mohtadi, and P. Tuffs. Generalized predictive control–Part I. The basic algorithm. *Automatica*, 23(2):137–148, 1987a.

D. Clarke, C. Mohtadi, and P. Tuffs. Generalized predictive control–Part II. Extensions and interpretations. *Automatica*, 23(2):149–160, 1987b.

C. R. Cutler and D. L. Ramaker. Dynamic matrix control–A computer control algorithm. In *AIChE 86th National Meeting*, Houston, TX, 1979.

M. L. Darby and M. Nikolaou. MPC: Current practice and challenges. *Control Engineering Practice*, 20(4):328–342, 2012. Special Section: IFAC Symposium on Advanced Control of Chemical Processes – ADCHEM 2009.

D. de la Peña, A. Bemporad, and T. Alamo. Stochastic programming applied to model predictive control. In *Proceedings of the IEEE Conference on Decision and Control and European Control Conference*, pages 1361–1366, 2005.

L. Deori, S. Garatti, and M. Prandini. Computational approaches to robust model predictive control: A comparative analysis. In *Proceedings of the IFAC World Congress*, pages 10820–10825, Cape Town, 2014.

S. Di Cairano. Indirect adaptive model predictive control for linear systems with polytopic uncertainty. In *Proceedings of the American Control Conference*, pages 3570–3575, Boston, MA, 2016.

J. L. Doob. *Stochastic Processes*. John Wiley & Sons, 1990.

S. E. Dreyfus. *Dynamic programming and the calculus of variations*. Academic Press, 1965.

M. Farina and R. Scattolini. Model predictive control of linear systems with multiplicative unbounded uncertainty and chance constraints. *Automatica*, 70: 258–265, 2016.

M. Farina, L. Giulioni, and R. Scattolini. Stochastic linear model predictive control with chance constraints – A review. *Journal of Process Control*, 44:53–67, 2016.

A. A. Feldbaum. Dual-control theory I. *Automation and Remote Control*, 21:874–880, 1961a. Translated from Avtomatlka i Telemekhanika, 21(9), 1960.

A. A. Feldbaum. Dual-control theory II. *Automation and Remote Control*, 21:1033–1039, 1961b. Translated from Avtomatlka i Telemekhanika, 21(9), 1960.

E. Fogel and Y. Huang. On the value of information in system identification – Bounded noise case. *Automatica*, 18(2):229–238, 1982.

H. Genceli and M. Nikolaou. New approach to constrained predictive control with simultaneous model identification. *AIChE Journal*, 42(10):2857–2868, 1996.

R. Ghaemi, J. Sun, and I. Kolmanovsky. Less conservative robust control of constrained linear systems with bounded disturbances. In *Proceedings of the IEEE Conference on Decision and Control*, pages 983–988, 2008.

J. Goh and M. Sim. Distributionally robust optimization and its tractable approximations. *Operations Research*, 58(4-part-1):902–917, 2010.

P. J. Goulart, E. C. Kerrigan, and J. M. Maciejowski. Optimization over state feedback policies for robust control with constraints. *Automatica*, 42(4):523–533, 2006.

M. Green and J. B. Moore. Persistence of excitation in linear systems. *Systems & Control Letters*, 7(5):351–360, 1986.

G. Grimm, M. Messina, S. Tuna, and A. Teel. Model predictive control: For want of a local control Lyapunov function, all is not lost. *IEEE Transactions on Automatic Control*, 50(5):546–558, 2005.

G. Grimm, M. Messina, S. Tuna, and A. Teel. Nominally robust model predictive control with state constraints. *IEEE Transactions on Automatic Control*, 52(10): 1856–1870, 2007.

L. Grüne and J. Pannek. *Nonlinear Model Predictive Control*. Communications and Control Engineering. Springer, Cham, Switzerland, 2nd edition, 2017.

L. Grüne and A. Rantzer. On the infinite horizon performance of receding horizon controllers. *IEEE Transactions on Automatic Control*, 53(9):2100–2111, 2008.

M. Guay, V. Adetola, and D. DeHaan. *Robust and Adaptive Model Predictive Control of Nonlinear Systems*. Control, Robotics and Sensors. Institution of Engineering and Technology, 2015.

B. Hassibi, A. H. Sayed, and T. Kailath. LMS is H^∞ optimal. In *Proceedings of the Conference on Decision and Control*, pages 74–79 vol.1, 1993.

T. A. N. Heirung, B. Foss, and B. E. Ydstie. MPC-based dual control with online experiment design. *Journal of Process Control*, 32:64–76, 2015.

T. A. N. Heirung, B. E. Ydstie, and B. Foss. Dual adaptive model predictive control. *Automatica*, 80:340–348, 2017.

M. Herceg, M. Kvasnica, C. N. Jones, and M. Morari. Multi-parametric toolbox 3.0. In *Proceedings of the European Control Conference*, pages 502–510, Zürich, Switzerland, 2013.

L. Hewing and M. N. Zeilinger. Cautious model predictive control using gaussian process regression. *ArXiv e-prints*, abs/1705.10702, 2017.

L. Hewing and M. N. Zeilinger. Stochastic model predictive control for linear systems using probabilistic reachable sets. *ArXiv e-prints*, abs/1805.07145, 2018.

S. Kanev and M. Verhaegen. Robustly asymptotically stable finite-horizon MPC. *Automatica*, 42(12):2189–2194, 2006.

E. C. Kerrigan. *Robust Constraint Satisfaction: Invariant Sets and Predictive Control.* PhD thesis, University of Cambridge, 2000.

T. H. Kim and T. Sugie. Adaptive receding horizon predictive control for constrained discrete-time linear systems with parameter uncertainties. *International Journal of Control*, 81(1):62–73, 2008.

A. Klenke. *Probability Theory: A Comprehensive Course.* Universitext. Springer London, 2014.

E. D. Klenske, M. N. Zeilinger, B. Schölkopf, and P. Hennig. Gaussian process-based predictive control for periodic error correction. *IEEE Transactions on Control Systems Technology*, 24(1):110–121, 2016.

I. Kolmanovsky and E. G. Gilbert. Theory and computation of disturbance invariant sets for discrete-time linear systems. *Mathematical Problems in Engineering*, 4(4): 317–367, 1998.

M. Korda, R. Gondhalekar, J. Cigler, and F. Oldewurtel. Strongly feasible stochastic model predictive control. In *Proceedings of the IEEE Conference on Decision and Control and European Control Conference*, pages 1245–1251, 2011.

M. Korda, R. Gondhalekar, F. Oldewurtel, and C. N. Jones. Stochastic MPC framework for controlling the average constraint violation. *IEEE Transactions on Automatic Control*, 59(7):1706–1721, 2014.

M. V. Kothare, V. Balakrishnan, and M. Morari. Robust constrained model predictive control using linear matrix inequalities. *Automatica*, 32(10):1361–1379, 1996.

B. Kouvaritakis and M. Cannon. *Model Predictive Control: Classical, Robust and Stochastic.* Springer International Publishing, Cham, Switzerland, 2016.

B. Kouvaritakis, M. Cannon, S. V. Raković, and Q. Cheng. Explicit use of probabilistic distributions in linear predictive control. *Automatica*, 46(10):1719–1724, 2010.

F. Kozin. A survey of stability of stochastic systems. *Automatica*, 5(1):95–112, 1969.

H. J. Kushner. *Stochastic Stability and Control*. Academic Press, New York, USA, 1967.

H. J. Kushner. A partial history of the early development of continuous-time nonlinear stochastic systems theory. *Automatica*, 50(2):303–334, 2014.

W. Langson, I. Chryssochoos, S. V. Raković, and D. Mayne. Robust model predictive control using tubes. *Automatica*, 40(1):125–133, 2004.

D. Limon, T. Alamo, D. M. Raimondo, D. M. noz de la Peña, J. M. Bravo, A. Ferramosca, and E. F. Camacho. Input-to-state stability: A unifying framework for robust model predictive control. In *Nonlinear Model Predictive Control Towards New Challenging Applications*, volume 384. Springer, 2009.

M. Lorenzen, F. Allgöwer, F. Dabbene, and R. Tempo. Scenario-based stochastic MPC with guaranteed recursive feasibility. In *Proceedings of the IEEE Conference on Decision and Control*, pages 4958–4963, Osaka, Japan, 2015a.

M. Lorenzen, F. Allgöwer, F. Dabbene, and R. Tempo. An improved constraint-tightening approach for stochastic MPC. In *Proceedings of the American Control Conference*, pages 944–949, Chicago, Illinois, 2015b.

M. Lorenzen, F. Allgöwer, and M. Cannon. Adaptive model predictive control with robust constraint satisfaction. In *Proceedings of the IFAC World Congress*, pages 3368–3373, Toulouse, France, 2017a.

M. Lorenzen, F. Dabbene, R. Tempo, and F. Allgöwer. Constraint-tightening and stability in stochastic model predictive control. *IEEE Transactions on Automatic Control*, 62(7):3165–3177, 2017b.

M. Lorenzen, F. Dabbene, R. Tempo, and F. Allgöwer. Stochastic MPC with offline uncertainty sampling. *Automatica*, 81:176–183, 2017c.

M. Lorenzen, M. A. Müller, and F. Allgöwer. Stochastic model predictive control without terminal constraints. *International Journal of Robust and Nonlinear Control*, 2017d.

M. Lorenzen, M. A. Müller, and F. Allgöwer. Stabilizing stochastic MPC without terminal constraints. In *Proceedings of the American Control Conference*, pages 5636–5641, Seattle, Washington, 2017e.

M. Lorenzen, M. Cannon, and F. Allgöwer. Robust MPC with recursive model update. *Automatica*, 2018. under review.

S. Lucia, T. Finkler, and S. Engell. Multi-stage nonlinear model predictive control applied to a semi-batch polymerization reactor under uncertainty. *Journal of Process Control*, 23(9):1306–1319, 2013.

L. Magni, H. Nijmeijer, and A. Van Der Schaft. A receding-horizon approach to the nonlinear $\mathcal{H}_\infty$ control problem. *Automatica*, 37(3):429–435, 2001.

M. Mammarella, E. Capello, F. Dabbene, and G. Guglieri. Sample-based SMPC for tracking control of fixed-wing UAV. *IEEE Control Systems Letters*, 2(4):611–616, 2018a.

M. Mammarella, M. Lorenzen, E. Capello, H. Park, F. Dabbene, G. Guglieri, M. Romano, and F. Allgower. An offline-sampling SMPC framework with application to automated space maneuvers. *IEEE Transactions on Control Systems Technology*, 2018b. under review.

G. Marafioti, R. R. Bitmead, and M. Hovd. Persistently exciting model predictive control. *International Journal of Adaptive Control and Signal Processing*, 28(6):536–552, 2014.

D. Mayne. Robust and stochastic model predictive control: Are we going in the right direction? *Annual Reviews in Control*, 41:184–192, 2016.

D. Mayne, J. Rawlings, C. Rao, and P. Scokaert. Constrained model predictive control: Stability and optimality. *Automatica*, 36(6):789–814, 2000.

D. Mayne, M. Seron, and S. V. Raković. Robust model predictive control of constrained linear systems with bounded disturbances. *Automatica*, 41(2):219–224, 2005.

D. Q. Mayne. Model predictive control: Recent developments and future promise. *Automatica*, 50(12):2967–2986, 2014.

A. Mesbah. Stochastic model predictive control: An overview and perspectives for future research. *IEEE Control Systems Magazine*, 36(6):30–44, 2016.

A. Mesbah, S. Streif, R. Findeisen, and R. D. Braatz. Stochastic nonlinear model predictive control with probabilistic constraints. In *Proceedings of the American Control Conference*, pages 2413–2419, Portland, Oregon, USA, 2014.

S. P. Meyn and R. L. Tweedie. *Markov chains and stochastic stability*. Cambridge University Press, Cambridge, UK, 2009.

H. Michalska and D. Q. Mayne. Robust receding horizon control of constrained nonlinear systems. *IEEE Transactions on Automatic Control*, 38(11):1623–1633, 1993.

J. Moore. Persistence of excitation in extended least squares. *IEEE Transactions on Automatic Control*, 28(1):60–68, 1983.

K. S. Narendra and A. M. Annaswamy. *Stable adaptive systems*. Dover Publications, 2005.

M. Ono and B. Williams. Iterative risk allocation: A new approach to robust model predictive control with a joint chance constraint. In *Proceedings of the IEEE Conference on Decision and Control*, pages 3427–3432, 2008.

C. J. Ostafew, A. P. Schoellig, and T. D. Barfoot. Conservative to confident: Treating uncertainty robustly within learning-based control. In *Proceedings of the IEEE International Conference on Robotics and Automation*, pages 421–427, 2015.

J. A. Paulson, A. Mesbah, S. Streif, R. Findeisen, and R. D. Braatz. Fast stochastic model predictive control of high-dimensional systems. In *Proceedings of the IEEE Conference on Decision and Control*, 2014.

L. S. Pontryagin. Optimal regulation processes. *Uspekhi Mat. Nauk*, 14(1):3–20, 1959.

M. Prandini, S. Garatti, and J. Lygeros. A randomized approach to stochastic model predictive control. In *Proceedings of the 51st IEEE Conference on Decision and Control*, pages 7315–7320, 2012.

A. Prékopa. *Stochastic Programming*, volume 324 of *Mathematics and Its Applications*. Springer Netherlands, 2010.

J. A. Primbs and C. H. Sung. Stochastic receding horizon control of constrained linear systems with state and control multiplicative noise. *IEEE Transactions on Automatic Control*, 54(2):221–230, 2009.

A. I. Propoi. Application of linear programming methods for the synthesis of automatic sampled-data systems [russian]. *Avtomatika i Telemekhanika*, 24(7): 912–920, 1963.

S. Qin and T. A. Badgwell. A survey of industrial model predictive control technology. *Control Engineering Practice*, 11(7):733–764, 2003.

S. V. Raković and Q. Cheng. Homothetic tube MPC for constrained linear difference inclusions. In *Proceedings of the 25th Chinese Control and Decision Conference*, pages 754–761, 2013.

S. V. Raković, B. Kouvaritakis, R. Findeisen, and M. Cannon. Homothetic tube model predictive control. *Automatica*, 48(8):1631–1638, 2012.

K. J. Åström and B. Wittenmark. *Adaptive Control*. Dover Publications, 2nd edition, 2008.

J. B. Rawlings, D. Bonne, J. B. Jorgensen, A. N. Venkat, and S. B. Jorgensen. Unreachable setpoints in model predictive control. *IEEE Transactions on Automatic Control*, 53(9):2209–2215, 2008.

J. B. Rawlings, D. Q. Mayne, and M. M. Diehl. *Model Predictive Control Theory and Design*. Nob Hill Publishing, Madison, Wisconsin, USA, 2nd edition, 2017.

M. Reble. *Model Predictive Control for Nonlinear Continuous-Time Systems with and without Time-Delays*. PhD thesis, Universität Stuttgart, 2013.

J. Richalet, A. Rault, J. Testud, and J. Papon. Model predictive heuristic control: Applications to industrial processes. *Automatica*, 14(5):413–428, 1978.

G. Schildbach, L. Fagiano, C. Frei, and M. Morari. The scenario approach for stochastic model predictive control with bounds on closed-loop constraint violations. *Automatica*, 50(12):3009–3018, 2014.

A. T. Schwarm and M. Nikolaou. Chance-constrained model predictive control. *AIChE Journal*, 45(8):1743–1752, 1999.

F. Scibilia, S. Olaru, and M. Hovd. On feasible sets for MPC and their approximations. *Automatica*, 47(1):133–139, 2011.

M. A. Sehr and R. R. Bitmead. Stochastic output-feedback model predictive control. *Automatica*, 94:315–323, 2018.

J. Skaf and S. Boyd. Nonlinear Q-Design for convex stochastic control. *IEEE Transactions on Automatic Control*, 54(10):2426–2430, 2009.

M. Tanaskovic, L. Fagiano, R. Smith, and M. Morari. Adaptive receding horizon control for constrained MIMO systems. *Automatica*, 50(12):3019–3029, 2014.

M. Tanaskovic, D. Sturzenegger, R. Smith, and M. Morari. Robust adaptive model predictive building climate control. In *Proceedings of the IFAC World Congress*, pages 1871–1876, Toulouse, France, 2017.

R. Tempo, G. Calafiore, and F. Dabbene. *Randomized Algorithms for Analysis and Control of Uncertain Systems: With Applications*. Springer, 2013.

D. van Hessem. *Stochastic inequality constrained closed-loop model predictive control with application to chemical process operation*. PhD thesis, Delft University of Technology, 2004.

D. van Hessem and O. Bosgra. Stochastic closed-loop model predictive control of continuous nonlinear chemical processes. *Journal of Process Control*, 16(3):225–241, 2006.

D. van Hessem, C. Scherer, and O. Bosgra. LMI-based closed-loop economic optimization of stochastic process operation under state and input constraints. In *Proceedings of the IEEE Conference on Decision and Control*, volume 5, pages 4228–4233, 2001.

S. M. Veres. Adaptive control by worst-case duality. In C. Bányász, editor, *5th IFAC Symposium on Adaptive Systems in Control and Signal Processing*, IFAC Postprint Volume, pages 73–78. Pergamon, Oxford, 1995.

S. M. Veres, H. Messaoud, and J. P. Norton. Limited-complexity model-unfalsifying adaptive tracking-control. *International Journal of Control*, 72(15):1417–1426, 1999.

S. Yan, P. Goulart, and M. Cannon. Stochastic model predictive control with discounted probabilistic constraints. In *Proceedings of the European Control Conference*, 2018.

X. Zhang, K. Margellos, P. Goulart, and J. Lygeros. Stochastic model predictive control using a combination of randomized and robust optimization. In *Proceedings of the IEEE Conference on Decision and Control*, Florence, Italy, 2013.

X. Zhang, S. Grammatico, G. Schildbach, P. Goulart, and J. Lygeros. On the sample size of randomized MPC for chance-constrained systems with application to building climate control. In *Proceedings of the European Control Conference*, pages 478–783, 2014.

Y. Zhang, D. Monder, and J. F. Forbes. Real-time optimization under parametric uncertainty: a probability constrained approach. *Journal of Process Control*, 12(3): 373–389, 2002.